MARITIME
STRATEGY
OR
COALITION
DEFENSE?

ROBERT W. KOMER

UNIVERSITY
PRESS OF
AMERICA

LANHAM • NEW YORK • LONDON

ABT
BOOKS

University Press of America,™ Inc.

4720 Boston Way
Lanham, MD 20706

3 Henrietta Street
London WC2E 8LU England

Library of Congress Cataloging in Publication Data

Komer, R. W.
    Maritime strategy or coalition defense?

    Reprint. Originally published: Cambridge, Mass. : Abt
Books, c1984.
    Includes index.
    1. Sea-power—United States. 2. World politics—20th
century. 3. United States—Military policy. I. Title.
[UA23.K746   1984b]   359'.03'0973   84-13137
    ISBN 0-8191-4117-8
    ISBN 0-8191-4118-6 (pbk.)

To My Father

Contents

Foreword

by **McGeorge Bundy**

One of the side effects of our necessary concern with the dangers presented by nuclear weapons is that we tend to neglect the large problems of choice presented in the rest of our defense programs. Such neglect is a mistake on at least four grounds. First, conventional forces account for the lion's share of our defense spending — four dollars out of five at a good rough guess. Second, the historical record shows that our conventional capabilities are put to frequent operational test, both in deployments for deterrence and in open combat. Third, it is increasingly agreed among our leaders, political and military alike, that conventional strength is a vital element in the deterrence of nuclear war itself, and for several years over two administrations there has been solid bipartisan support for substantial increases in conventional spending. Fourth, we are now confronted by an unavoidable requirement for hard choices in the use of limited resources. There is no political leader in either party who will support everything the separate services now claim to require, and unless hard choices are made by deliberate decision, they will be arbitrarily forced by budgetary constraints. This book is about such choices, and it is the best yet written on the subject.

Robert Komer's thesis is that up to now we have emphasized inherited service preferences over serious strategic choices, and nationalistic bad habits over the requirements of interdependence with friendly countries. He finds our most serious present error in costly overemphasis on new carrier battle groups, and he traces that error back to both a faulty strategic purpose — one that might be called containment by maritime counterforce — and a faulty assessment of risks — one that puts undue emphasis on the possibility of large-scale conventional maritime warfare in many places at once. For Komer the proper center of conventional effort is the maintenance of a balanced level of strength, in all arms, that can allow us to play an effective part in the defense of the places on land that matter most — especially Western Europe, Japan, and the oil-rich Middle East. He finds that while maritime strength is absolutely essential for keeping the sea-lanes open, a maritime strategy alone is bound to fail, and he uses both history and present geopolitical reality to make his case.

Finding the central requirement that of effectively deterrent capabilities on land, Komer goes on to argue the necessity for cooperation with allies and friends. Writing from a wide and deep experience of politico-military affairs, he argues that coalition defense, with all its difficulties, is both cheaper and better than any effort to go it alone, that its demands for both American resources and American leadership are not being met, and that it now faces further grievous weakening from budgetary constraints unless the present unbalanced emphasis on forces of lesser value is reversed.

I think Komer's answers are essentially right, not only on their own strategic terms but on the wider ground that they are responsive to political imperatives as well. His position is squarely in the great tradition of our postwar military political leadership, all the way back to Marshall and Eisenhower. I believe that only by what Komer calls a coalition strategy can we help to sustain the political will and self-confidence that are essential to the future safety of all the societies we need as friends.

Even more fundamental than right answers are right questions. When an unavoidable choice in public policy is presented, it is deeply wrong that it should escape public examination. As one who has known and respected Robert Komer for over twenty years, I salute this further demonstration of his exemplary instinct for the jugular.

Preface

This book is about *nonnuclear* strategy and posture — an assessment of U.S. conventional strategic options in an age of nuclear stalemate. Given some thirty-five years of dealing professionally with politico-military and operational issues, mostly as a U.S. official, I write not as an academic theorist or armchair strategist but as an active participant in the national security process in Washington and in the field. My approach is pragmatic rather than theoretical, focusing on real strategic issues that we currently confront in the conventional force arena.

Why focus on nonnuclear strategy? First and foremost because the advent of nuclear stalemate has created a difficult period of transition from primary reliance on nuclear deterrence to much greater reliance on nonnuclear deterrence and defense. This alone compels us to rethink our strategy and forge a new consensus on it within the Western alliance. However, this book does not argue for abandonment of nuclear deterrence, an unrealistic goal in foreseeable circumstances, only that overreliance on nuclear weapons can be a fatal strategic flaw. It is hard to fault the view of Andrei Sakharov that, catastrophic as use of these awesome weapons would be, sheer prudence requires that we maintain the balance of terror until we can tame the nuclear monster via arms controls.[1]

Second, at a time when the military balance has been turning against us even more in the conventional military arena than in the nuclear, it is imperative that we make realistic strategic choices if we can't do everything. As Senator Sam Nunn of Georgia keeps reminding us, the United States is simply stretched too thin. Hence the practical definition of strategy used herein is *the art of making realistic choices in a context of constrained resources* — relating ends to means. This is hardly an elegant definition (there are many different levels of strategy), but any student of past wars or veteran of the Pentagon budget battles will know what I mean.

Third, a rich literature is already available on nuclear strategy, long the subject of a great deal of attention. U.S. thinkers have clearly brought out how profoundly U.S. strategy was affected by the advent of the nuclear age, compelling us to shift toward relying primarily on a strategy of deterrence. Nor have nuclear war-fighting strategies been neglected. In fact, U.S. nuclear strategy has been in continuous evolution — particularly after we foresaw the end of massive U.S. nuclear superiority and shifted to flexible response and then to more sophisticated nuclear deterrent and war-fighting doctrines developed by the last three administrations.

Nonnuclear strategic issues, however, have rarely received comparable analysis in depth, not even when adoption of "flexible response" in the 1960s signaled their increased importance. With some notable exceptions like Samuel Huntington, William Kaufmann and Edward Luttwak, few U.S. military professionals or civilian defense intellectuals have written much about nonnuclear strategy. The very term strategic has been corrupted to refer only to strategic nuclear issues. True, we Americans often talk about policy, program, doctrine, or even tactics as "strategy." For example, victory through air power or command of the sea seem to me more strategic doctrines than actual strategies, however much our military services tend to dress them up in the latter guise. We even occasionally ornament our arguments with strategic principles culled from such as Clausewitz or Mahan.

Rarely, however, do we Americans address ourselves imaginatively to the art of how best to achieve broad military objectives with the capabilities in hand at any given time. Of course, our unified and specified commands routinely prepare a variety of contingency plans, but from some knowledge of these I would regard most of them as rather unimaginative, with inadequate consideration of alternatives. Instead we tend to focus on technology, weapons systems, or force structure without relating them to strategic missions. "The impetuous pace of technology . . . has outdistanced strategic thought," as Sir James Cable says in lamenting a similar lack of focus in recent British thinking. He finds government inhibitions to be so great that such thinking has to come from outside.[2]

In my view this lack creates a serious gap in our thinking about national security affairs. Logically, the process of generating our defense posture should start by defining our strategic interests (vital or

otherwise), assessing strategic alternatives for protecting these interests, and only then proceeding to generate capabilities designed to
execute the strategy chosen. In reality, of course, the process is far
more complex, especially for a global superpower that must plan and
posture against a wide range of contingencies.

In fact, I am tempted to ask whether many administrations, not
just the current one, really have had much of a strategy beyond some
broad constructs such as deterring Soviet aggression, containing Soviet expansionism, honoring our commitments or staying flexible. In
any case, preferred strategy seems to have had only limited influence
on our conventional defense posture compared to many other factors
which necessarily shape it in real life. Rather than being geared to
our strategic priorities, our force posture tends to be dictated more
by political factors, economic constraints, and parochial competition
for constrained resources among the various military services. Indeed, Richard Betts argues that this will inevitably be the case in a
society like ours. He finds defense budget ceilings "determined by
vague senses of threat and acceptable risk, not by strategy."[3] In ana-
lyzing U.S. Navy budgets, Abellera and Clark expound the thesis that
"old habits of national security" are more likely to determine the
navy's future budgets than "the Navy's vision"of how it must develop
to support national strategy.[4]

The traditional reluctance of democracies to spend adequately
on defense in peacetime is another powerful influence (see Chapter
2). Perhaps Congress is most guilty on this score. The capabilities we
have already generated greatly affect what strategy we can realistically
expect to support in the near term. Our commitments also affect our
strategy, even though yesterday's commitments may no longer correspond with today's strategic interests (as in the case of Taiwan).
Therefore, I have not hesitated to discuss these relevant issues in a
book on the conventional strategic options open to us today.

We also tend to neglect the way the objective of deterrence has
given strategy an even larger peacetime dimension than it had before.
For example, achieving credible deterrence has heightened the dis-
tinction between *declaratory* strategy — what we threaten to do in
event of conflict — and *operational* strategy, what we would actually
try to do if war came. The increased importance of *perceptions* in a
nuclear age means that we must think a lot more carefully about
declaratory strategy as well. Operating on an adversary's perceptions
of what he could and could not reasonably expect to accomplish is
even more crucial in an age of mass destruction weaponry than it
was before.

Now that the United States is so dependent on allies and is compelled to pursue a coalition policy, we must also heed the sage advice

of Michael Howard that any sound coalition strategy must not only influence enemy perceptions but reassure our allies.[5] I too have long been a vigorous exponent of the need for a coalition strategy and posture, dictated by the fact that the United States must rely ever more heavily on its network of alliances to maintain a satisfactory balance of power vis-a-vis the Soviet bloc. As I develop in this book, and plan to elaborate in a larger work on *Coalition Defense* for the same publisher, we can no longer afford what General David Jones has called "the sin of unilateralism."

This leads to the main purpose of this book. At long last, nuclear stalemate and resultant rising inhibitions over nuclear war-fighting are leading to a renaissance of thinking about conventional strategy (see Chapter 4), to which I hope this book will contribute. We need to think a great deal more seriously about what we would do in case of nonnuclear conflict — particularly now that the United States no longer has that great cushion of strategic nuclear superiority to shield it from perceived need for much conventional strategic thought. Nor is it any longer enough to gear our posture to the "worst case" and assume that this will suffice for most lesser contingencies as well. Hence our security interests dictate rethinking our nonnuclear strategy and posture.

Even so, military men in particular may ask why a civilian is tackling this subject. For a short answer, I will cite the distinguished Frenchman who claimed that there are two fields in which amateurs seem to do as well as professionals — prostitution and strategy. A more serious response is that our military professionals have not done much serious study of conventional strategic issues. Moreover, the U.S. military establishment finds it institutionally very difficult to render unified strategic advice. Instead, our civilian leadership often has to choose among at least four separate strands of military service thinking — army, navy, air force, and marine — because the Joint Chiefs of Staff (JCS) have proved systemically incapable of resolving the deep differences among them. This means that the strategy advocated by the JCS tends to be the sum of all the strategic desires of these services.

Thus when the JCS complain about the mismatch between our strategy and our resources, what they mean is that the budgets provided are insufficient to enable us to execute all the services' strategic (or doctrinal) aims. When I agreed with General Jones, then chairman of the JCS, about the mismatch but argued that we should modify our strategy to the extent we could not find adequate resources, he told me candidly that it would be almost impossible to get the Joint Chiefs to do so. His frustration, and that of Army Chief of Staff General E. C. Meyer, over their inability to get the JCS to focus on such key issues, helped lead both of them to become ardent public advocates of JCS reform — a need that I heartily endorse.[6]

A word about my own credentials for offering some informal contributions on strategy and force posture might be appropriate at this point. In the 1950s I helped prepare National Intelligence Estimates for the top policymakers, including those on Soviet military capabilities. Subsequently, seven years detailed to the Eisenhower National Security Council (NSC) staff, then as a member of the Kennedy/Johnson NSC staff, and finally as deputy special assistant to the president for national security affairs gave me a splendid window on the policy process. Duty in Vietnam, where I tried to generate a strategic approach oriented to pacification, added to my experience. As ambassador to Turkey in 1968/1969 I served as U.S. permanent representative to the Central Treaty Organization Council, which produced insights on coalition strategy. During 1969/1976, Rand Corporation teams led by me produced three major studies on how to rejuvenate NATO's conventional option, which provided much of the foundation for the Carter administration's NATO initiatives.

But it was my recent Department of Defense experience during 1977/1980 that both catalyzed my thinking and offered opportunities to carry it into practice. After designing the U.S. initiatives launched at the May 1977 NATO summit, including proposed development of a new Long-Term Defense Program, I initiated and followed through on a broad range of measures to rejuvenate NATO's defense posture as Secretary Brown's Advisor on NATO affairs. Then as Undersecretary of Defense for Policy in 1979/1980, I served as the titular chief civilian policy and strategic planner in the Pentagon. Among my responsibilities was preparing the *Defense Policy Guidance* (DPG) issued annually by the secretary to guide program planning. Believing that policy and strategy should indeed guide programming, I invested a great deal of effort in framing the 1979 and 1980 DPGs along these lines — even attempting to set regional and functional priorities.

Another strong conviction was that strategic mobility was the Achilles' heel of our force projection strategy and posture, in that we possessed more ready forces in the United States itself than we could deploy rapidly enough to meet contingency needs.[7] General E. C. Meyer and I did successfully advocate the rapid reinforcement concept of fielding a "ten-division D-Day force"in Europe within ten to fourteen days, via prepositioning the equipment for six reinforcing divisions. I also became an ardent advocate of airlift and fast sea lift.

What finally put strategic mobility back on the map in 1980 was the new contingency requirement for rapid deployment to the Persian Gulf. Indeed, my appointment as Undersecretary coincided with the Southwest Asia crisis of 1979/1980, when the United States perceived an urgent need to develop quick response capabilities in yet a third major theater — the Persian Gulf. Having had some experience with this area (back in 1963 I was the one who proposed seeking from Britain a joint base in the Indian Ocean, which led to Diego Garcia), I spent much of 1980 trying to help think through how best to put credible flesh on the bones of the Carter Doctrine — a practical case in relating force posture to strategy. In retrospect, the demanding requirements of the Persian Gulf scenario have been perhaps the most important stimulus to U.S. reappraisal of the need for lighter, more flexible U.S. ground and tactical air forces, and greater strategic mobility.

With JCS acquiescence, Secretary Brown also made me the first Pentagon civilian official assigned on his behalf to review military contingency planning assumptions and final contingency plans to ensure they were consistent with U.S. policy. I drafted the first *Planning Guidance for Contingency Planning*. All this gave me more familiarity than most civilians with our military contingency planning. In addition, three years chairing DoD's Mobilization Steering Group, set up after a major mobilization exercise (*Nifty Nugget* in 1977) showed me how poorly prepared we were to go to war. This exercise, and the later *Proud Spirit* exercise of 1980, confirmed that we still have an outmoded mobilization doctrine inherited from World War II, which does not correspond adequately to future strategic needs. I became the chief exponent of greater allied host nation support for our projection forces, and personally convinced Congress to increase from 50,000 to 100,000 the number of reservists we could call up without declaring a state of national emergency.

Grappling with such actual problems sharpened my perception that we need to rethink our conventional strategy, not just provide more resources. Teaching two university seminars on "Strategic Issues of the 1980s," I was also struck by the parallels between the issues the United States confronts today and similar problems faced earlier by other maritime powers, especially England and Japan (see Chapter 5). Both nations recurrently addressed whether to rely primarily on maritime power as their chief line of defense or whether their vital interests demanded sending an army to the continent as well. In Britain this led to a "centuries-long debate between what have been termed the 'maritime' and the 'continental' schools of strategy."[8]

As will be seen, I regard these as the two main contending real-life schools of nonnuclear U.S. strategy too, especially under the current administration. The tension between them is not merely a theoretical issue. However imperfectly articulated, it has had, and is

having, a major impact on how we allocate the endemically constrained resources that free societies are willing to spend on peacetime
defense. This is a trillion dollar issue over time, and how we address
it could have a great deal to do with how well we emerge from the
continuing U.S.-Soviet military competition.

My deep concern is that the United States might be drifting in
practice toward an unbalanced maritime strategy and posture at the
expense of other equally pressing needs, and that this will not only
undermine our alliances but will lead to disaster in case of a major
nonnuclear war. This concern was first laid out in a *Washington Post*
article in February 1982, and then in a more extended treatment in
Foreign Affairs (Summer 1982). This book, which has the same title
as the latter, represents a further development and expansion of the
issues I raised then. Let me also emphasize that "maritime versus
continental" strategy is not an either-or proposition, but a question
of priority emphasis. Although my views will inevitably be taken as
having an antinavy cast, I fully recognize that command of the sea
at times and places of our choosing is vital to the overseas force-
projection strategy wisely practiced by the United States. Brian Bond's
conclusion in *British Military Policy Between the Two World Wars* that
British security demanded *both* control of the sea and a continental
commitment is equally applicable to the United States — and is a
major thesis of this book.

What I criticize herein is rather the natural institutional pro-
pensity of navies to argue for strategies and programs that emphasize
maritime striking power, regardless of the cost to optimally balanced
forces when resources are constrained. I am attacking not the need
for a powerful U.S. Navy but for a particular type of very costly navy,
one which stresses maritime force projection even at the expense of
sea control. Nor am I attacking aircraft carriers when I argue that
twelve modern carrier battle groups are the most we can afford when
other pressing priority needs are taken into account — including
other naval needs (see Chapter 6). Readers will draw their own judgment as to the validity of my case.

Finally, my gratitude to all those who have instructed me, often
without knowing it, such as Professors Michael Howard, Paul Ken-
nedy, and Brian Bond, Generals David Jones, E. C. Meyer, P. X.
Kelly, Richard Lawson, and Paul Gorman. Especial thanks to Harold
Brown, who as Secretary of Defense encouraged me to think about
such issues and was always willing to back a strategic proposition
whenever I could convince him (far from always) that I knew what
I was talking about. My thanks as well to Tom Brown, Paul Davis,
Kevin Lewis, and John Mearsheimer for their very helpful critiques.
Lastly, I am immensely grateful to The Rand Corporation for letting

me use its facilities while I wrote this book, and above all for letting Susanne Farmer assist me in producing it. Without her I could not have done it.

R. W. Komer
Washington, D.C.

NOTES

1. Andrei Sakharov, "The Danger of Thermonuclear War, An Open Letter to Sidney Drell," *Foreign Affairs*, Summer 1983, pp. 1001/1016.

2. James Cable, *Britain's Naval Future*,(London: Naval Institute press, 1983), pp. 2, 8.

3. R. K. Betts, "Conventional Strategy: New Critics, Old Choices,"*International Security*, Spring 1983, p. 148.

4. James Abellera and Rolf Clark, "Forces of Habit: Budgeting for Tomorrow's Fleets," *AEI Foreign Policy and Defense Review*, Vol. 3, Nos. 2, 3 (1981), p. 2.

5. Michael Howard, "Reassurance and Deterrence," *Foreign Affairs*, Winter 1982/83, pp. 309/324.

6. See General David Jones and Gregory Foster, *Battles Lost on the Potomac*, (Cambridge, Mass.: Abt Books, forthcoming).

7. For the most incisive analysis of the imbalance between our force posture and our strategic mobility capabilities, see W. W. Kaufmann, *Planning Conventional Forces 1950/1980*, (Washington, D.C., Brookings Institution, 1982), pp. 16/17.

8. Paul M. Kennedy, *The Rise and Fall of British Naval Mastery*, (New York: Scribner's, 1976), p. xvi.

CHAPTER 1

Evolution of U.S. Strategy

It is worth beginning any discussion of U.S. strategy with a brief review of how it evolved, because one key reason Americans seem to neglect strategy is that for most of our history it did not pose much of a problem. "In neglecting the subject, American thinkers were only following a traditional path. Grand strategy for a whole war . . . had never been an object of serious study in the United States. The geographical position of the nation and its relations to other countries seemed to make the development of a national strategy unnecessary."[1]

Once our fledgling republic found its feet, it pursued for well over its first hundred years a policy of "splendid isolation," heeding the advice of our first president to avoid entangling alliances.[2] This policy was squarely based on an enormous geopolitical advantage — that the United States was protected from hostile pressure or involvement in great power quarrels by two broad moats, the Atlantic and Pacific oceans. It was also greatly facilitated both by a stable balance of power in Europe and by Britain's mastery of the seas throughout the nineteenth and early twentieth centuries. Indeed, our Monroe Doctrine of 1823 warning off the European powers from attempting to reestablish colonial influence in the Americas was really underwritten not by the tiny U.S. Navy but by Britain's Royal Navy — in pursuit of a parallel British policy.

Not that the young republic was devoid of expansionist impulses. These, however, found ample scope for realization in the westward

movement across an almost empty continent. After two abortive attempts to take over Canada in its early years, the United States absorbed northern Mexico in the Mexican War. Then, as the nation
recovered from the Civil War and began looking outward, from about
1880 on it logically gave priority to building a respectable navy. Following the precepts of Mahan, the focus after 1890 was on building
a battleship fleet, though without much of a strategic rationale. Nonetheless, the creation of a naval war college in 1884 and the teachings
of Luce and Mahan gave the navy a clearer strategic focus than the
army. Throughout the nineteenth century our miniscule peacetime
army was largely employed in Indian wars or in coast defense of key
ports.[3]

Maintaining the Balance of Power via Force Projection

Although the Spanish-American War of 1898 saw the beginnings of
a force-projection strategy and acquisition of overseas bases, World
War I marked the real U.S. entry into balance-of-power politics. From
a strategic viewpoint, we entered the war in 1917 to help prevent
imperial Germany from upsetting the balance of power and, not
incidentally, to help protect our growing overseas trade. Both the
navy and army were expanded massively. In 1916 President Wilson
called for building a navy "bigger" than Great Britain's, and Congress
voted even more than the navy asked for in the Naval Act of 1916.
But the Grand Fleet had already bottled up the German fleet, and
the battleship building program had to be suspended in favor of
building antisubmarine vessels. In any case, World War I offered
little scope for a U.S. strategy because we were essentially tied to that
already adopted by our allies. To this end we sent a large expeditionary force to Europe in 1918.

World War I thus confirmed U.S. adoption of what might be
called a *force-projection* strategy. If the chief determinants of a satisfactory global balance of power lay on the other side of the ocean,
then U.S. attempts to maintain that balance necessarily entailed sending large forces overseas to aid embattled allies. Force projection had
another great advantage. Since we fought our wars on someone else's
real estate, the destructiveness of war was confined to them. The
United States to all intents and purposes was left untouched. Hence
force projection became a fixed element of our strategy, greatly in-
fluencing our force posture and eventually our peacetime deployments.

After World War I the United States withdrew again into isola-
tionism and even passed laws such as the 1934 Neutrality Act de-

signed to forestall another such intervention. Again we cut back our army to a shadow, forcing it to plan for another massive mobilization if we again had to project large forces overseas. The navy fared rather better. War with Japan, which would naturally be mainly a maritime conflict, became the principal contingency against which it planned. Within the limits imposed by the Washington and London naval treaties, we maintained a powerful fleet, which absorbed the lion's share of peacetime defense spending.[4]

Although Japan's surprise attack at Pearl Harbor launched us into World War II, the underlying cause was the inexorable need to preserve a satisfactory balance of power. World War II marked the definitive entry of the United States into balance-of-power politics. It also witnessed the first big arguments between our army and our navy, over the balance of resources to be allotted to the Pacific (mostly a navy theater, although General MacArthur made the same case) versus European theaters. The fact that European strategy was a U.S.-U.K. coalition affair, plus Churchill's influence over Roosevelt, gave the army's Europe-first case added clout. On the other hand, U.S. and British indecision over how soon to invade the continent from Britain gave the navy repeated opportunities to siphon off resources for the Pacific. "As late as September 1943 there were as many Army divisions opposing Japan as there were opposing Germany." If one added in navy and marines, the Pacific then was still by a small margin the area of greater U.S. manpower commitment.[5] The Europe versus Pacific debate, however, was not really an argument between maritime and continental strategies, since geography dictated that in both cases sea power was essential to seizing key land objectives. Indeed, the chief competition between the European and Pacific theaters was for all-important landing craft.

Deterrence via Alliances Backed by Massive Retaliation

The perceived Soviet and Chinese threat in the immediate post-World War II period and the decline in European (especially British) capabilities to cope with it led the United States to reverse its long-standing aversion to peacetime alliances in favor of a globe-girdling network — NATO, then SEATO, CENTO, and ANZUS — designed to help contain Soviet expansionism in the so-called Cold War. Along with these alliances went two other new practices — semipermanent forward deployments of ground/air as well as naval forces on the soil of key allies and a large foreign military and economic assistance program — that both enhanced deterrence and facilitated U.S. force projection in event of conflict.[6] All this served a "grand strategy" of containment, which the United States has pursued with remarkable consistency ever since the emergence of a perceived Soviet threat soon after World War II.[7]

Moreover, World War II spurred a technological revolution — the advent of nuclear weapons and intercontinental delivery means — that had even more profound strategic implications for the United States. For the first time, North America itself was no longer protected by two broad oceans from the risk of devastating attack. The U.S. homeland itself, like that of the other superpower, became highly vulnerable.

This in turn dictated another major departure from traditional U.S. security posture. Up to this point the United States had tended to starve its armed forces in peacetime and then, while its allies held the ring, belatedly mobilize and arm huge forces once peacetime unpreparedness had helped lead to war. Now its growing vulnerability to direct nuclear attack compelled the United States to adopt *a strategic aim of deterrence* — backed up by comparatively large ready peacetime forces designed primarily to deter any foe from launching a major conflict in the first place.[8] These large peacetime forces were also needed for our peacetime strategy of containment, as was expansion of the program of military aid begun in 1947.[9] This grand strategy was laid out in NSC-68, the first definitive statement of U.S. national strategy ever adopted.[10] Nonetheless, U.S. defense budgets fell precipitately in 1946/1950. Only $13 billion in outlays was proposed for FY 1951 until Kim Il Sung's surprise attack on South Korea.

The issue of strategic priorities arose again in the Korean War. Because of precipitate postwar U.S. demobilization, the JCS had declared South Korea outside our strategic perimeter, thus helping to invite North Korea's June 1950 surprise attack.[11] The JCS saw this as potentially a secondary-theater diversion at the expense of defense of Europe. MacArthur's rash advance to the Yalu, which brought China into the war, led JCS Chairman General Bradley to dub it "the wrong war, at the wrong place, at the wrong time, and against the wrong enemy." Nonetheless, the U.S. strategic focus shifted to Asia with the Korean and first Indochina wars, and the two Taiwan Straits crises of 1954 and 1959.[12]

The costs and frustrations of the Korean War, combined with the traditional reluctance of democracies to spend much on preparedness in peacetime, contributed to U.S. adoption of a strategy of primary deterrent reliance first on its nuclear monopoly and then on its massive nuclear superiority. Though individually very costly, nuclear weapons and their delivery systems were far less expensive than the large conventional forces otherwise deemed necessary to cope with those of the USSR. In short, they permitted defense on the cheap.

The Korean War also had an unanticipated side effect of great strategic significance. It triggered a major U.S. rearmament program whose impetus lasted until the Vietnam War, thus putting teeth into NSC-68's call for rectifying the deterrent balance of power.[13] Although nuclear deterrence continued to dominate U.S. strategic thinking and defense posture, NSC-68 and then the Korean War also led to an abortive effort to build up conventional forces, based on the NSC-68 estimate that "within four years the Soviet Union would have enough atomic bombs and a sufficient capability of delivering them to offset substantially the deterrent capability of American nuclear weapons."[14] Thus as far back as early 1950, U.S. decision makers foresaw the advent of a nuclear stalemate that would both enhance the utility of the Soviet advantage in conventional ground forces and create a danger of limited communist military adventures designed to remain below the nuclear threshold.

Events developed largely as foreseen. The threat of nuclear retaliation did not suffice to deter less than superpower conflicts. Korea and Vietnam, the two major wars in which the United States has engaged since 1945, were both with lesser powers helped by the U.S.S.R. and China. Hence the United States developed an implicit strategic doctrine of limited war, mostly unstated because it would detract from nuclear deterrence. The United States would respond conventionally to limited aggression where it perceived such a necessity, but it would be careful not to escalate hostilities (as in MacArthur's march to the Yalu) to the point where serious risks of a larger confrontation would arise.

Extended nuclear deterrence remained our primary strategic instrument in Europe, however, after our NATO allies boggled at the 1952 Lisbon goals of some 96 divisions and 9000 aircraft proposed by the NATO military authorities as the minimum essential for conventional defense. Our European allies, still recovering from World War II, found nuclear deterrence to be defense on the cheap for them too, since the United States — for its own nationalistic reasons — was willing to foot the great bulk of the nuclear deterrent bill. Indeed, the Eisenhower administration's adoption of its "massive retaliation" strategy, even to deter or fight limited wars, was partly in order to hold down defense spending.[15] Conventional forces, especially ground forces, were reduced. NATO too adopted this U.S. strategy in MC 14/2 of 1956. It was facilitated by U.S. development and deployment of tactical nuclear weapons and by the advent of new jet bomber and then missile delivery systems. "But to seek refuge in technology from hard problems of strategy and policy was . . .

another dangerous American tendency, fostered by the pragmatic qualities of the American character and by the complexity of nuclear age technology."[16]

Emergence of Flexible Response Strategy

As the U.S.S.R. seemed to be catching up, the new Kennedy administration, concerned (too much so, as it later turned out) about new Soviet ICBM programs, shifted U.S. strategy to one of so-called *flexible response*.[17] Instead of relying so heavily on an immediate and crushing nuclear response, the United States would build up its conventional forces to permit an initially conventional response — a pause during which NATO could hold its own while deciding on nuclear retaliation and confronting the aggressor with this risk. Kennedy and Johnson began the rebuilding of U.S. conventional forces. Defense outlays rose significantly from $45.7 billion in FY 1960 to $54.1 billion by FY 1964, including Atomic Energy Commission (AEC) expenditures).[18]

However, U.S. attempts to get NATO to agree to this revised strategy encountered great allied reluctance, not least to what seemed to be a weakening of U.S. resolve to retaliate massively with nuclear weapons (thus continuing to provide Europe with cheap deterrence) as opposed to the costly conventional buildup that flexible response implied. "Flexible response" was first proposed to NATO in May 1962, but not until December 1967 did NATO officially adopt it.[19] Even then it was only in a deliberately ambiguous compromise formulation (MC 14/3), which permitted the allies to interpret the strategy as calling for only a brief conventional "pause," with only a modestly strengthened tripwire to trigger nuclear escalation, whereas the Americans favored building toward an indefinite conventional defense. Meanwhile U.S. attention was diverted by the Vietnam War.

After its long entanglement in Southeast Asia, the United States again turned to trying to shore up its neglected European flank. In 1973/1975 Secretary of Defense Schlesinger returned to the fray, arguing vigorously for a more "stalwart" NATO conventional defense, especially since the USSR's nuclear buildup was gradually undermining the credibility of U.S. reliance on nuclear retaliation. However, Schlesinger found our allies still reluctant to pay the price. True, they modestly increased their own defense outlays by 20 percent during 1970/1976 (while U.S. outlays declined 13 percent as we phased out of Vietnam and then cut back further in expiation). These allied increases, however, went mostly for modernization of existing forces, not increases in force structure.

Then the Carter administration picked up the cudgels, arresting the decline in U.S. defense spending and beginning a gradual increase. Its far-reaching NATO initiatives, mostly launched at a U.S.-proposed NATO summit in May 1977 (little over three months after the new president took office) and then approved at another summit only one year later, were aimed primarily at coopting our allies into a sustained nonnuclear buildup. The centerpiece, a new Long-Term Defense Program, was to be funded by an agreed 3 percent real annual growth in defense spending for the next several years.

The United States simultaneously responded to growing European concern over the adverse "Euro-strategic balance," in Helmut Schmidt's phrase, being created by Soviet SS-20 missile and Backfire bomber deployments at a time of strategic nuclear stalemate. But this long-range theater nuclear force modernization program was not the centerpiece of the U.S. NATO initiatives. It was more an add-on designed to reassure our allies that no gap in the deterrent spectrum would be allowed to develop while we all focused on thickening NATO's conventional shield. Even 572 missiles (mostly ground-launched cruise missiles-GLCMs) were not regarded as providing NATO with a militarily adequate escalatory option. Besides, GLCMs (with relatively long flight times) are not much of a first-strike system; instead, they serve primarily to prevent the U.S.S.R. from holding Europe a nuclear hostage.

Nor was the United States ever enthusiastic about the buildup of British and French nuclear forces. We regarded them as superfluous, potentially destabilizing (could we rely on de Gaulle not to pull the nuclear trigger?), and almost inevitably funded at the expense of U.K. and French conventional forces. Although we aided U.K. nuclear force modernization — most recently in promising to provide Trident II missiles — this was essentially because we saw the British as determined to modernize anyway, hence calculated that we might as well help them do so less expensively in order to minimize the impact on their conventional NATO contribution.

True, the Pentagon has gradually developed more sophisticated nuclear war-fighting strategies designed not only to enhance deterrence but to facilitate escalation control if deterrence failsd and either side crosses the nuclear firebreak. National Security Decision Memorandum 242 of 1974, developed under the aegis of Secretary Schlesinger, called for a variety of more discriminating nuclear options rather than reliance on a single spasm response. The next step in the process was the "countervailing strategy" promulgated in Presidential Determination (PD) #59 of 1980.[20] National Security Decision Directive #13 of 1982 is essentially an updating of PD #59. Though these evolutionary developments reflect a desire to prolong the utility of nuclear deterrence *in extremis*, the dominant thrust of every U.S. administration since Kennedy's has been to strengthen conventional forces to compensate for what was nevertheless perceived as the inevitably declining credibility of extended nuclear deterrence as the Soviets caught up in nuclear capabilities.

In short, *it is we Americans who have been trying for over twenty years to get our allies to move away from dangerous overreliance on nuclear weapons.* It is allied governments that have clung to the U.S. nuclear crutch, primarily because of their reluctance to pay for adequate conventional forces. Thus it is ironic to find a large segment of European opinion accusing the United States of wanting to fight a theater nuclear war at Europe's expense.

Similarly, the United States has almost always taken the lead in efforts to help tame the nuclear monster via arms control. The apparent interruption of this process — when SALT II could not be ratified and then the Reagan administration deliberately held back so that the United States could negotiate later from a "position of strength" — turned out to be a tactical error, which the administration is now seeking to rectify.

Meanwhile, it is important to remember that even a U.S. nuclear monopoly and then massive superiority did not deter limited conventional conflicts — of which the largest were Korea and then Vietnam. Almost from the outset, nuclear powers were self-deterred from using the awesome weapons they had invented. The U.S.S.R. even carefully avoided any direct confrontation with the United States, instead encouraging surrogates to exploit perceived opportunities. It is significant that not since 1919 (in Siberia and around Archangel) have Soviet and U.S. forces directly engaged each other. Indeed, not until the invasion of Afghanistan at the end of 1979 did the U.S.S.R. ever use its own forces in combat beyond the World War II boundaries of its empire.

The United States in turn was careful to keep local conflicts limited, allowing sanctuaries to its opponents and receiving sanctuaries in return. Thus, while nuclear deterrence was by no means absolute, it did make the superpowers cautious about keeping limited war limited, especially when their competitive interests were directly involved. In this sense it seems that nuclear deterrence has worked. Despite many prophecies of doom, there has been no major superpower war in the thirty-eight years since the end of World War II. Nor have basic U.S. security interests yet been compromised.

Nevertheless, how much longer can this favorable situation be expected to last? To cite one wise and knowledgeable observer:

For the last twenty years an uneasy feeling has been spreading throughout this land that the United States is losing its extraordinary primacy in the global arena; that it is laboring under a multitude of external and internal constraints which increasingly frustrate its efforts to support its ever-expanding global security interests. Developments in the 1970s have greatly reinforced this pervasive sense of constraint — particularly, the shift of the military balance toward the Soviet Union, the divergence of U.S. and European views on East-West relations following the rise and fall of detente, the rising turbulence in the Third World, the collapse of the international economic system, the greatly enhanced dependence of the great industrial states on Middle Eastern oil, and finally the impact of the Vietnam War on American will and means to back global containment with military power. In the aggregate these constraints raise with new urgency an old concern that American security responsibilities and commitments may exceed American power to support them. Whether this concern is completely warranted or not, there are objective bases for it. To assess the bases for the concern that the ends of American foreign policy may have outrun the means, and to formulate the broad strategic consequences, are tasks of preeminent importance. They are tasks for *grand strategy,* which is the nation's plan for using all its instruments and resources of power to support its interests most effectively.[21]

NOTES

1.   T. Harry Williams, *History of American Wars*, (New York: Knopf, 1981), pp. 194/195.

2.   Washington's advice, however, ignored the fact that our own revolution probably could not have succeeded without the help of our French ally — not just in sending a fleet and troops, but also in keeping Britain preoccupied with much larger wartime problems than one of several colonial sideshows. France under Napoleon provided much the same service to us in the War of 1812.

3. See Russell Weigley, *The American Way of War*, (New York: Macmillan, 1973), pp. 188/191, for a critique of preWorld War I naval thinking. See also Williams, *History of American Wars*, pp. 306/308.

4.   Weigley, *American Way of War*, pp. 242/246.

5.   Ibid., p. 271.

6.   On the merits of forward deployment and their low cost, see R. W. Komer, "Future of U.S. Conventional Forces: A Coalition Approach," in *Rethinking Defense and Conventional Forces,* (Washington, D.C.: Center for National Policy, 1983), p. 47.

7.   Robert E. Osgood, "American Grand Strategy: Patterns, Problems,

and Prescriptions," *Naval War College Review*, September/October 1983, p. 7.

8.   Weigley, *American Way of War*, pp. 365/367.

9.   This was not adopted without controversy. See Dean Acheson, *Present At The Creation*, (New York: Macmillan, 1969), pp. 307/309.

10.   The best treatment of NSC-68 is Paul Hammond, "NSC-68: Prologue to Rearmament" in *Strategy, Policy, and Defence Budgets,* (New York: Columbia University Press, 1962), pp. 271/378. Later iterations of NSC-68 came to be called "Basic National Security Policy." However, the document became so complex and the subject of such interagency nitpicking that the attempt to update it was finally abandoned in the early Kennedy administration (on the author's recommendation, among others).

11.   Although the JCS had provided this advice, largely on grounds that postwar defense cuts made it imperative to limit our commitments, it was left to Secretary of State Acheson to announce the decision.

12.   In 1955, after the French defeat in Indochina, the JCS declared that Indochina was "devoid of strategic interest." Leslie Gelb, "When Is a Foreign Interest 'Vital'?," *New York Times*, August 8, 1983, p. 14.

13.   Weigley, *Amarican Way of War*, pp. 380/381.

14.   W. W. Kaufmann, *Planning Conventional Forces 1950-1980,* (Washington, D.C.: Brookings Institution, 1982), p. 2.

15.   Secretary of State Dulles is cited to this effect in Robert McNamara, "The Military Role of Nuclear Weapons," *Foreign Affairs*, Fall 1983, p. 62.

16.   Weigley, *American Way of War*, p. 416.

17.   In *Doubletalk: The Story of SALT I*, (New York, Doubleday, 1980), pp. 10/11 Ambassador Gerard Smith describes how as early as 1958 he convinced Secretary of State Dulles that "massive retaliation" was becoming outdated, and convinced the Secretary to so advise the JCS.

18.   Weigley, *American Way of War*, p. 447. This buildup later became distorted by the exigencies of the Vietnam War.

19.   McNamara, "The Military Role of Nuclear Weapons," *Foreign Affairs*, Fall 1983, pp. 63/64.

20.   See Harold Brown, *Thinking About National Security*, (Boulder, Colo.: Westview Press, 1983), pp. 77/85 for a cogent analysis of the rationale behind this strategic evolution.

21.   Osgood, "American Grand Strategy," p. 5.

CHAPTER **2**

Why We Must Revamp U.S. Strategy

Turning to the military aspects of grand strategy, the province of this book, Osgood's thesis that our security interests exceed our present capability to support them seems amply justified. In particular, neither the United States nor its allies have yet faced up to the full conventional force implications of continuing over-reliance on nuclear deterrence or generated sufficient effort to put more flexibility into flexible response. On the contrary the current Supreme Allied Commander Europe (SACEUR), General Bernard Rogers, has complained repeatedly of being condemned to "inflexible response" because NATO's conventional capabilities are so inadequate as to require him to ask for nuclear escalation very quickly.[1] Many civilian critics of our nuclear posture do pay lip service to the need for conventional alternatives, but seldom have they presented any realistic calculations of what would be required to this end — or how much it would cost. Yet many factors have cumulatively led to a now urgent need to rethink our military strategy and posture.[2]

Advent of Nuclear Stalemate

Probably the most important factor has been the arrival of nuclear stalemate. Its political impact, along with related factors, has also created "a serious disjunction between deterrence and reassurance" - the twin objectives of NATO policy.[3] One result has been a rising popular reaction in the West to the seemingly inexorable piling up of advanced nuclear weapons on both sides, and an often emotional search for ways out of the nuclear dilemma. It has led to proposals for a U.S. no-first- use pledge, various forms of nuclear freeze, build-down of nuclear arsenals, or even unilateral disarmament. At bottom these are symptoms of an underlying fact of life — *that the advent of nuclear stalemate makes a strategy based primarily on nuclear retaliation less and less appealing to the very people it is designed to protect.*[4]

The military implications of nuclear stalemate are equally striking. As former Secretary of Defense James Schlesinger put them recently, "in the past United States strategic nuclear force superiority offset weaknesses . . . in our conventional force structure. . . . Our basic concern today is that our strategic forces can no longer compensate for deficiencies in our conventional forces."[5] U.S. efforts to prolong the utility of nuclear deterrence via more sophisticated nuclear strategies do not alter this basic reality. In the last analysis, it is increasingly being perceived that nuclear weapons mainly deter use of other nuclear weapons.

This is not to say that the extension of the U.S. nuclear umbrella over our allies has lost all credibility. It will remain a powerful deterrent — above all to direct superpower confrontation — if only because the risks of nuclear escalation would rise greatly were the United States and the U.S.S.R. already engaged in conventional war. Accident or miscalculation would become much more likely, as would the risk that the side that perceived itself to be losing might be sorely tempted to escalate. It is precisely for this reason that stronger conventional capabilities are badly needed as deterrent insurance, despite their higher cost.

Decline in Relative U.S. Power

The U.S.S.R.'s catching up in nuclear capabilities is only one aspect of a broader decline in U.S. power — vis-a-vis the U.S.S.R. in a military sense, but more in relation to U.S. allies and the Third World in terms of economic strength. The sharpest decline has been in the U.S. share of world economic output. As late as 1949 we produced over half the world's gross national product (GNP). In the early 1980s we are producing only about one-quarter. This is not because the U.S. economy has become poorer. On the contrary, we are wealthier than ever. Rather, the economies of other nations have grown faster than ours — in particular those of Western Europe and Japan as they recovered (with generous U.S. aid) from the ravages of World War II. The Third World's share of world output has also grown significantly, partly because of the sharp rise in oil prices. Thus, although the U.S.S.R. is still far from achieving Khrushchev's goal of burying the United States economically, there has been a real diffusion of economic power.

Shift in Military Balance

But in terms of pure military power it is the U.S.S.R. which has caught up with and in some respects surpassed the United States. "As far as American strategy and national security policy are concerned," Huntington finds this "the most important change of the 1970s."[6] After Washington faced down Moscow in the 1962 Cuba missile crisis, the U.S.S.R. embarked on a sustained military buildup, increasing its real annual military outlays by a variously estimated 2/4 percent each year for the next twenty years. Of course, the only way the U.S.S.R. can generate such high military spending, investing some 14/16 percent of GNP from an economy much smaller than ours, is to give top priority to its military buildup — in strong contrast to democratic societies. Significantly, the U.S.S.R. has also been able to generate a much higher proportion of defense investment from its military outlays, primarily because its proportional "people costs" remain much lower than those of the United States. Today, with our volunteer system, almost half our defense outlays still go for people costs (including retirement), whereas the Soviet proportion is estimated at more like 25 percent. As a result the gap in all-important defense investment (research and development, procurement, and construction) has become even wider. Moreover, the relative stability of Soviet defense spending contrasts sharply with the swings in U.S. defense outlays.

While Soviet defense spending was going up, U.S. defense spending was going sharply down, until 1977. The prime cause was hardly U. S."neglect," as often claimed, but the costs and consequences of the Vietnam War. In FY 1984 dollars, Vietnam cost us some $331 billion, an enormous diversion of defense effort. Then in the aftermath we expiated the Vietnam disaster by further cutting real defense outlays another $130 billion in FY 1984 dollars between 1972 and 1977.[7] In short, the direct and indirect costs of Vietnam amounted to an enormous diversion of defense dollars — even leaving aside the erosive effects of the rampant inflation unleashed largely by the way the United States financed the Vietnam War and of the higher pay to volunteers when Vietnam led to the demise of the draft.

This combination of increased Soviet military effort and a real decline in the U.S. effort has gradually altered the military balance. Indeed, if there is such a thing as a "window of vulnerability," it lies much more in the conventional disparity between Western and Soviet bloc forces than in any nuclear disparity. Aside from catching up in strategic nuclear capabilities and outstripping NATO in long-range theater missiles, the U.S.S.R. has greatly increased its ability to project conventional military power. In fact, like the United States, it has spent far more on the latter than the former.

As a result, the U.S.S.R.'s capability to attack conventionally in the NATO, China/Japan/Korea, or Persian Gulf theater has grown more rapidly than have Western defensive capabilities. In the crucial European theater NATO has at best "only a delayed tripwire" defense.[8] Also striking is the growth in Soviet ability to intervene in remote areas as a result of the buildup of its navy and airborne forces and its active search for bases and surrogates overseas. Although U.S. force-projection capabilities remain far superior in areas beyond Eurasia, they face increasing challenge. This deterioration of the U.S.-Soviet military balance has not escaped our perceptive allies. According to one thoughtful Japanese diplomat:

> How to restructure . . . cooperation for Western Security acutely concerns Western industrialized democracies, including Japan. The shift of the U.S.-Soviet military balance from one of American supremacy to one of ambiguous equilibrium, with its profound implications for the security perceptions of the U.S. and her allies, is the most fundamental though not the sole development arousing such concerns.[9]

Reluctance of Democracies To Spend Adequately on Defense

Underlying the shift in the military balance is the traditional reluctance of democratic societies to provide adequately for defense in peacetime, no matter how often history shows that deterrence is far cheaper than war.[10] Modern British history is replete with such examples; while World War II and Korea provide ample U.S. illustrations. Yet we Americans somehow fail to take into account the horrendous costs, both human and financial, of inadequate peacetime preparedness. Moreover, deterrence is not just cheaper than war; it has become a strategic imperative in a nuclear age.

Nonetheless, democratic societies, with their pluralistic goals, still tend to make deterrence/defense almost a residual call on national resources.[11] In the 1960s and 1970s social spending had clear priority and increased multifold in most developed nations, while defense declined in terms of the percentage of an expanding GNP it absorbed. For example, the U.S. Congress, which more closely reflects popular attitudes than does the presidency, cut the president's defense budget requests every year for fifteen years, until FY 1980.[12]

Add to this the even greater reluctance of democratic governments to spend on defense in times of economic difficulty, another characteristic of which examples abound. The recent global recession and sharply higher inflation triggered by the redoubling of oil prices in 1979 had precisely this effect. They were perhaps the chief reasons why the European NATO allies failed to meet the 3 percent annual real growth goal for military spending adopted by their own heads of government only a year earlier. In the event, only the United States achieved at least 3 percent real growth from 1978 to the present.

This tendency to hold down defense spending, as opposed to the way the Soviet command economy can steadily generate a high and expanding level by fiat, has had a distinct impact on the military balance. It has also led Western governments to seek defense on the cheap by concentrating on what appear to be decisive solutions. For example, the Eisenhower administration chose massive retaliation as its preferred strategy in the 1950s, largely because this permitted holding down defense costs. A major reason that our allies also favored nuclear deterrence is that it was even cheaper for them, since the United States paid most of the bills. As Betts says, "it is only realistic to recognize that domestic politics determines military options far more than expert strategic analysis does."[13]

Expansion of U.S. Strategic Requirements

Largely because of resource constraints, the United States has never generated postwar conventional capabilities on the scale the Joint Chiefs of Staff considered necessary for potential multifront contingencies. The JCS have usually expressed this link between conventional strategy and posture in terms of worst-case "force-sizing" scenarios in which U.S. forces might have to fight in several theaters at the same time, or at least keep on guard in some while fighting in others. When General Bradley called Korea "the wrong war at the wrong place and the wrong time," he reflected a JCS consensus that North Korea's attack (and China's subsequent intervention) might be only a diversion facilitating renewed Soviet pressures in Western Europe, with which the precipitately demobilized postwar U.S. forces would be wholly unable to cope. Kaufmann describes how later adoption of a "massive retaliation" strategy sidetracked the conventional force buildup called for by NSC-68.[14]

The next attempt to develop a conventional force-sizing scenario came with adoption of a "flexible-response" strategy in the early 1960s. A 1962 study by the military planners apparently looked at about sixteen separate potential theaters of conflict and concluded that U.S. forces would be needed to back up allied forces in eleven of these more or less simultaneously. However, the resulting requirement was so huge, and the likelihood so limited without prior extensive Soviet mobilization, that the planners cut it back to the so-called two-and-a-half-war scenario — a simultaneous Warsaw Pact attack on NATO, Chinese communist campaign in Korea or Southeast Asia, and lesser contingency somewhere else. Even this was estimated to require some twenty-eight and a third U.S. divisions and forty tactical air wings, active and reserve, in addition to allied forces.[15] These plans and the ambitious buildup envisaged were knocked into a cocked hat by the exigencies of the Vietnam War.

As the Nixon administration wound down U.S. participation in Vietnam and took advantage of the Sino-Soviet split to develop contacts with China, it was clear that U.S. defense efforts would be cut back — on top of the costly impact of the Vietnam War itself on the U.S. defense posture. The administration's 1969 shift to a one-and-a-half war force-sizing concept, after a careful assessment of several alternative force postures, was almost certainly a shrewd adjustment to these realities. It was another attempt to cut the coat to fit the cloth.

The Carter administration also started with a force-sizing study pursuant to Presidential Review Memorandum #10. Few significant increases in conventional force structure were sought, although modernization was pursued. Nor was there any significant shift of contingency force allocations to the European theater from other theaters. On the contrary, it was a new commitment looming on the horizon that forced the Carter administration to rethink the one and one-half war scenario.

Although the U.S. military had long made contingency plans for the Persian Gulf-Indian Ocean region, we long regarded it as primarily a U.K. responsibility. As Britain gradually withdrew from east of Suez, London and Washington sought to fill the gap by creating an alliance with the northern tier states — Turkey, Iran, and Pakistan — whose large forces could be modernized through military aid. This CENTO alliance served a useful deterrent purpose. It began to crumble as Pakistan shifted toward a pro-Chinese orientation and the 1974 Cyprus crisis led to a U.S. embargo on military aid to Turkey. But the quadrupling of OPEC oil prices in 1974 enabled the shah of Iran to attempt to fill the gap himself using his newfound oil wealth.

The fall of the shah and Soviet invasion of Afghanistan (which seemed to betoken a new expansionism) led President Carter to reaffirm the U.S. deterrent umbrella over this newly vulnerable oil-rich area. His Carter Doctrine of January 1980 announced that the United States would resist, by force if necessary, any Soviet or other attempt to take over the oil of the Persian Gulf. This compelled the United States to adopt what I called at the time a "one and two-half war" force sizing scenario (a major war in Europe plus smaller conflicts in Korea and the Persian Gulf).

During its last year in office, the Carter administration created a new Rapid Deployment Force (RDF), mostly from existing assets, and began planning in terms of a requirement for five to seven army and marine divisions and seven air wings to deter or halt any Soviet move.[16] We simultaneously put our NATO and Japanese allies on notice that defending Persian Gulf oil that was more essential to them than to us would necessarily involve some diversion of U.S. forces from NATO or Pacific defense. We called on them to make compensatory increases in their own home defense capabilities.[17]

Mismatch Between Strategy and Resources

All these factors — the relative decline of U.S. power, the way in which nuclear stalemate has increased the need for costly conventional forces, the gradual shift in the superpower military balance, Western reluctance to spend on defense, and finally the perceived need to take more seriously a new Persian Gulf threat have led the JCS in particular to complain about the growing mismatch between U.S. strategic requirements and the resources available to meet them. Indeed, Robert Osgood finds that a chronic "interests-power gap" between our expanding security interests and our ability to support them has been a persisting problem since World War II. He sees postwar U.S. foreign policy "as largely shaped and driven by repeated efforts to close the gap between ever-expanding security interests and persistently inadequate power to support them, if the hypothetical threats to these interests should materialize."[18]

Although reluctance to spend adequately on defense was less of a disadvantage during the era of massive U.S. nuclear superiority, it has become more so as the advent of nuclear stalemate dictates greater reliance on expensive conventional forces. Constrained dollars lead to more intense service competition for them. The Reagan administration, to its credit, succeeded in increasing defense spending, although its search for a more coherent strategy to guide this spending has been less successful (see Chapter 6).

Contributing to the mismatch on the nonnuclear side is the continuing high cost of maintaining nuclear parity. Though this takes only an average of 15 percent of the first Reagan Five Year Defense Program, it takes a much higher proportion of the all-important defense investment segment, thus squeezing the resources left for conventional modernization. According to one estimate, the administration proposes to spend on the order of $450 billion in FY 1984 to FY 1989 on nuclear forces.

Greater Likelihood of Conflict

Yet another factor that argues for rethinking our strategy is that the likelihood of conflict directly engaging the interests of the super-powers seems greater in the 1980s than it was in the 1970s. This is not to contend that it has become *likely*, especially not in the sense of deliberate major superpower aggression, but rather that the in-hibitions to lesser conflicts are probably less than they were in the last decade. Above all, U.S. nuclear superiority, which undoubtedly had much to do with such Soviet caution in exploiting opportunities, has been superseded by an uneasy nuclear stalemate.

Whether this will actually lead to greater Soviet risk-taking is of course problematical. However, partially because it has given such high priority to a military buildup, the Soviet leadership finds itself in a systemic economic crisis, facing continued declining growth un-less it opens up the Soviet system, which in turn would threaten the leadership's control. The U.S.S.R. is hardly going to bury us eco-nomically, as Khrushchev claimed. At the same time, the ideological appeal of Soviet communism has declined sharply as other nations perceive that it does not offer what it claims. Under these circum-stances, there seems greater risk that the U.S.S.R. would be tempted in a crisis to use — or threaten to use as the backdrop for political pressure — the great military power it has accumulated, especially since Western military weakness and stresses within the Western al-liance may lead to a Soviet perception that the West could be faced down.[19] This has significant implications for our future defense pos-ture, not least the optimum balance between readiness and modern-ization.

In sum, the overriding security dilemma confronting U.S. poli-cymakers in the 1980s and beyond is how best to preserve credible deterrence and defense at a cost politically acceptable to democratic societies reluctant to divert too much to defense. This dilemma is sharpened by the objective of deterrence, which requires greater U.S. emphasis on ready peacetime forces, in contrast to the crisis or war-time mobilization responses characteristic of earlier wars. If the ad-vent of nuclear stalemate now dictates greater reliance on more expensive conventional forces, the dilemma posed becomes much more acute than in the past.

NOTES

1.   Interview with General Bernard Rogers, *Armed Forces Journal*, Sep-tember 1983, pp. 78/80.

2. Huntington describes "the widely shared belief" among partici-pants in his program on National Security Studies at Harvard's Center for International Affairs that "many of the key concepts that informed and shaped American strategy in the 1950s and 1960s no longer have the same relevance and usefulness in the 1980s." S. P. Huntington, ed., *The Strategic Imperative* (Cambridge, Mass.: Ballinger, 1982), p. ix.

3. Michael Howard, "Reassurance and Deterrence," *Foreign Affairs,* Winter 1982/1983, p. 317.

4. Robert W. Komer, "Maritime Strategy vs. Coalition Defense," *Foreign Affairs,* Summer 1982, pp. 1126/1127.

5. *U.S. News and World Report,* January 10, 1983, p. 19.

6. Huntington, ed., *The Strategic Imperative,* pp. 1/2.

7. Figures updated to FY 1984 constant dollars provided me by Program Analysis and Evaluation staff in the Office of the Secretary of Defense.

8. Rogers Interview, p. 80.

9. Yukio Satoh, "Western Security: A Japanese Point of View," *Naval War College Review,* September-October 1983, p. 75.

10. Huntington contests the prevalent myth that arms race buildups increase the likelihood of war, contending that the opposite is more often the case. See "Arms Races: Prerequisites and Results," *Public Policy* (Yearbook of Harvard Graduate School of Public Administration), 1958.

11. This argument is related to the again fashionable thesis that defense expenditures are at the expense of economic growth and stability. See Cable's dissection of this thesis in the British case in James Cable, *Britain's Naval Future* (London: Naval Institute Press, 1983), pp. 28/35.

12. R. James Woolsey, "The Defense Buildup: Just the Backlog Will Take Billions," *Washington Post,* February 17, 1981, p. 19.

13. R. K. Betts, "Conventional Strategy: New Critics, Old Choices," *International Security,* Spring 1983, p. 154.

14. W. W. Kaufmann, *Planning Conventional Forces 1950-1980* (Washington, D.C.: Brookings Institution, 1982), pp. 2/3.

15. See ibid., pp. 4/11, for Kaufmann's fascinating reconstruction of the strategic assumptions behind the two-and-a-half-war force-sizing scenario. Given the possibility that actual conflicts might occur in quite different locales from those planned for, the McNamara concept called for minimizing overseas deployments while building up in the continental United States a large strategic reserve with adequate air and sea lift to deploy it flexibly in any direction.

16. See Kaufmann, *Planning Conventional Forces,* p. 19 for the increases in strategic mobility planned. The original RDF requirement for 3/4 divisions and air wings had been sized to meet a non-Soviet threat, but this was soon recognized as inadequate to halt a Soviet thrust southward.

17. Undersecretary of State Newsom and I made the first official demarches to the NATO Council on April 14, 1980.

18. Robert E. Osgood, "American Grand Strategy: Patterns, Problems,

and Prescriptions," *Naval War College Review*, September-October 1983, pp. 5/7.

19.   When I suggested this likelihood in a memorandum on key strategic issues in late 1980, which was actively discussed in the NSC itself, no one challenged this hypothesis. Seweryn Bialer is similarly "drawn inescapably to the conclusion that . . . the years ahead . . . will witness the external expansion of an internally declining power." *After Brezhnev*, Georgetown University Center for Strategic and International Studies, Indiana University Press, p. 65.

CHAPTER **3**

Coalition Policy Without Coalition Posture

Up to this point U.S. defense policy and strategy have been treated mainly in terms of what the United States itself has done. This is too narrow a frame. We must also take into account both our allies and friends and those of the other superpower, who are central to the strategy we each pursue.

Coalition War as the Norm Rather Than Exception

Indeed, throughout recorded history most important conflicts were not fought by one country against another. They were usually fought between alliances, or coalitions. As an early example, the Peloponnesian War was a conflict not just between Athens and Sparta, but between the Delian League led by Athens and a Spartan-led coalition of city-states. Similar coalitions were a feature of the Punic Wars between Rome and Carthage. Shifting allies and alliances have been the norm rather than the exception in most major conflicts.

Even all four twentieth century U.S. wars were coalition wars. For World Wars I and II, this goes without saying. In Korea and Vietnam, although the United States was the senior partner, our South Korean and South Vietnamese allies fielded more forces and took more casualties than we did. In all four cases our allies provided the battlefields and suffered great damage, a not inconsiderable contribution to alliance burden-sharing (of which more later).

Yet, notwithstanding the strategic and operational problems repeatedly created by coalition warfare, it is amazing how little the defense postures and programs of modern nation-states have taken into account this reality. Surprisingly little doctrine, theory, or even analysis of these problems has emerged. This is as true of other major powers as it is of the United States. Their professional literature, including their official histories, scants this set of issues — except for some treatment of the endemic problems of combined command. The Russians, Germans, French, and British — despite far more bitter experience at coalition war than the United States has had — do not appear to have studied its problems any more seriously than we have. Instead, each nation seems to revert in peacetime to designing and posturing its forces as if it were going to fight the next war alone.[1]

This "sin of unilateralism," as General David Jones called it when he was Air Force Chief of Staff, reflects at bottom the powerful nationalism that still dominates politico-military thinking. It is inherent in the pluralistic nature of Western societies and reflects the primacy of nation-states. Postwar experience, my own included, also testifies to the numerous political, economic, commercial, and bureaucratic obstacles that inevitably inhibit coalition planning and programming, even in an alliance like NATO.

U.S. Alliance Policy

The sin of unilateralism persists despite the fact that the United States and most of Western Europe (and even, in time, their wartime enemies — Germany and Japan) soon after World War II adopted a firm alliance policy for collective security against a perceived communist threat. In the early days of NATO, all further felt that an integrated defense, via sharing defense burdens, would limit the cost to each participant. Dean Acheson recalls how on December 1, 1949, the NATO defense ministers agreed on an ingenious concept that no European ally was to attempt a complete military establishment, but rather "to make its most effective contribution in the light of its geographic position, economic capability, and population.[2] "Balanced collective forces instead of individually balanced national forces" was to be the order of the day. "Economic necessity dictated that all duplication be eliminated."[3] It would also help solve the problem of how best to incorporate an eventual German defense contribution.

Despite such early recognition that a rationalized collective effort was required, however, this alliance policy has never been matched by a comparable effort to generate a coherent alliance posture, In practice, rationalization was never carried out (except insofar as the

United States, for its own reasons, provided the bulk of NATO strategic nuclear and blue-water naval forces). Nationalism soon resumed its ascendancy.

Another key reason why defense integration soon lost its steam, since neither the allies nor the United States felt much need to build up costly conventional forces, was the American and then NATO shift to primary reliance on nuclear deterrence. Since Washington preferred to fund the nuclear umbrella over Europe, our still recovering European allies found the new U.S. nuclear strategy a convenient reason not to build up substantial conventional forces of their own. The abortive Lisbon Goals of 1952 (for fifty active and forty-six reserve divisions), which marked the last major attempt to generate much larger NATO conventional forces, were soon abandoned as infeasible. A similar fate awaited the European Defense Community, designed to be a truly integrated defense establishment and as such strongly supported by Washington.

True, NATO and its Soviet-led counterpart (the Warsaw Pact organization) are unique in the history of multinational alliances in the extent of their peacetime links. NATO has lasted thirty years and has created a combined command structure down to army group and tactical air force level, although the only forces actually assigned in peacetime are air defense units.[4] Combined programs include a commonly funded air defense radar system, high level communications network, integrated AWACS force, infrastructure program, and the like. NATO committees constantly strive to coordinate almost all facets of national programs. But what has actually been accomplished only scratches the surface of what would be possible if the NATO allies could summon up the political will (see Chapter 9). Moreover, the Soviets have prodded their Warsaw Pact allies into an integration process, including stress on standardization and interoperability, well beyond what NATO has achieved.

The United States has since made other efforts to persuade its allies to put more emphasis on conventional defense. Our attempt during the Kennedy/Johnson years to get NATO to adopt a "flexible response" strategy envisaged a much greater conventional effort. But little was forthcoming in the 1960s, partly because U.S. energies were soon diverted by the Vietnam War. NonU.S. NATO defense spending actually declined in real terms 1963/1969, while U.S. spending soared after 1965.[5] This comparison is highly misleading, however, because 1965/1972 U.S. defense outlays were dominated by the Vietnam War. Our actual NATO contribution in FY 1964/1972 markedly declined; indeed, we probably disinvested in NATO defense.

The reverse happened during the 1970s. Whereas in 1971 the United States still accounted for 61 percent of total NATO spending, this dropped to only 55 percent by 1974, owing mostly to the Vietnam and postVietnam drawdown. Indeed, allied outlays increased in real terms over 20 percent during 1971/1980, while U.S. outlays declined 11 percent.[6]

Perhaps the most comprehensive U.S. effort to get the NATO allies to do more, according to common plans and programs, came with the so-called Carter initiatives of 1977/1979. It was Carter administration insistence that led the allies to pledge 3 percent real growth per annum in May 1977. This was the backdrop for President Carter's far-reaching proposals at the May 1977 NATO summit in London for a set of "quick- fix" defense improvements, a challenge to NATO to design a new Long-Term Defense Program (LTDP), and a U.S. pledge to greater "alliance cooperation" in arms production, including acceptance of more of a two- way street in reciprocal defense sales. The LTDP, centerpiece of the Carter initiatives, was adopted at a second summit in Washington one year later, and the heads of government again pledged 3 percent real growth to fund it.

Though only the United States achieved an average 3 percent real growth in 1978/1980, arguably the Carter initiatives did lead to a higher level of allied defense spending than would otherwise have occurred.[7] By NATO definition, nonU.S. NATO allies did achieve an average real increase of 2.2 percent for 1979, 2.7 percent for 1980 and 2.7 percent for 1981.[8] But it was uphill going because of the recession triggered by the 1979 redoubling of oil prices, from which Europe has not yet recovered. Meanwhile, the Reagan administration achieved a further sharp rise in U.S. defense outlays, which again widened the gap between U.S. and allied spending.

Allied Burden-sharing

Even though allied defense efforts have generally fallen short of what NATO and U.S. authorities perceived as required, we Americans have consistently undervalued what they do contribute to the common defense. Indeed "the problem of equity is endemic in voluntary alliances."[9] American criticism of our allies has grown recently as we debate how much more the United States should spend on defense. It smacks of scapegoating.[10]

In fact, it is misleading to compare U.S. and allied defense efforts solely on the basis of respective percentages of GNP. We posture globally, while our allies (even Britain and France) posture the great bulk of their forces for a single regional contingency. Hence the agreed NATO method of using total defense outlays as the basis of comparison seriously understates the comparative European contribution to defense of Europe itself, a point made repeatedly by successive SACEURs, who keep pointing out that our allies provide the great bulk of the active forces currently deployed in NATO Europe.

Moreover, in certain categories, such as nuclear weapons, we Americans do not want the allies to spend. We prefer to minimize proliferation by providing the Free World nuclear shield ourselves. Since our geographic situation and global interests require a force-projection capability, we also spend a lot on airlift/sea lift and logistics that the allies do not. Besides, our expensive volunteer system results in far higher people costs than do continental European conscription systems. For example, DoD calculated that if 1979 allied manpower costs were computed at U.S. pay scales, "the value of nonU.S.-NATO total defense would increase relative to the United States by approximately 20 percent, reaching a total approximately equal to that of the United States."[11]

Nor can allied contributions be measured in terms of defense outlays alone. Our NATO allies maintain active forces of over 3 million men compared to our 2 million, and more than double the trained reserves. NATO allies account for over 55 percent of total NATO and Japanese ground combat capability, 50 percent of tactical air combat forces, and 35 percent of naval combatant tonnage, even when all U.S. forces (not just those deployed in Europe or earmarked for reinforcement) are taken into account. They also contribute an enormous amount of free real estate to our overseas forces (an estimated $80 billion worth by Germany alone), provide a wide range of peacetime host nation support (HNS), and plan to provide a far wider range of wartime HNS. They provide the bulk (effectively 80 percent) of NATO infrastructure funding.[12]

Last but not least, the realities of geography have meant — and would mean in the future — that *our NATO allies provide most of the battlefields in event of nonnuclear war*. The enormous cost in war damage and civilian casualties in World Wars I and II (or Korea and Vietnam) suggests how great a burden our allies again would carry in any future conventional conflict. If the United States finds it far more sensible strategically to fight its wars on foreign battlefields, surely this cost to our allies must be taken into account. All in all, as a perceptive British analyst argues, NATO would do well to shift from a concept of "equality of sacrifice" as the measure of burden-sharing to one of "equality of effort." This would enable NATO "to switch attention from bickering about how burdens are shared to working out how the labor might best be divided if NATO is to remain relevant and effective."[13]

Inadequate Alliance Cooperation

Concepts of equity aside, however, it remains indisputable that allied spending has been dangerously inadequate to meet the needs for even high-confidence *initial* defense (see Chapter 7). If conventional deterrence/defense in Europe or Northeast Asia seems so inadequate, this is far more attributable to the lack of European NATO and Japanese efforts than to lack of U.S. effort. Despite postwar reconstruction and sizable economic growth, they remain content to shelter beneath the U.S. nuclear umbrella and behind U.S. command of the seas. Countries like Japan, Canada, and Denmark make such modest defense efforts as to be open to accusations of getting a free ride by any standard. France and, to a lesser extent, Greece have even taken advantage of the U.S. umbrella to pursue independent policies that actually impede mutual defense.

Nor has the NATO alliance ever realized the potential for increased coalition efficiency feasible through common programs, specialization of effort, and more rational burden-sharing hoped for by its founders. "Balanced collective rather than balanced national forces" remains a distant goal. Although NATO's biennial Ministerial Guidance for Defense Planning has emphasized "Alliance Cooperation" ever since its innovative 1975 version, it has been honored more in the breach than in the observance. This is not to denigrate what has been accomplished, but even today NATO has probably realized only about 10/15 percent at best of the potential gains from working together more efficiently (not to mention those possible through integration).

At bottom NATO is still a collective facade behind which remains a collection of disparate national forces, with different tactics, procedures, force structures, logistic systems, and mostly nonstandardized or even noninteroperable equipment. NATO's "force goals" presented every two years for routine ministerial approval are mostly a collection of national plans modifiable only at the margin. Wasteful overlap and duplication in logistics, overhead, communications, and training establishments are the rule rather than the exception.

Nowhere is the waste greater than in R&D and procurement — at a time when inflation and increasing sophistication are making each new generation of equipment almost too costly to buy in the quantities needed. True, many NATO nations buy from each other, and there have been some useful coproduction schemes. The Anglo-German-Italian Tornado is a splendid aircraft, but its cost has gone out of sight.[14] The American-Belgian-Dutch-Norwegian-Danish F-16 consortium has been more cost-effective. At U.S. initiative, the United States, Britain, France, and Germany have jointly developed a new Multiple Rocket Launcher System and appropriate munitions, the first time the four biggest arms producers in NATO have jointly developed any major system (Italy has now joined too). Undersecretary of Defense William Perry championed an intensive effort at greater armaments cooperation, including such innovative concepts as the United States and Europe each taking the lead in developing one member of a "family of weapons" (and then each coproducing the other's designs).[15]

For the most part, however, NATO today still designs and fields a bewildering variety of diverse equipment for basically the same military missions. The urge to protect national technology bases, commercial advantage, and sheer jobs prevents much cooperation. Each nation's military bureaucracy is strongly resistant to change. Yet only in the United States (and in Germany for ground force equipment) is there sufficient national demand to permit economies of scale.

Allies Vital to U.S. Strategy

It is long since time for us to recognize that allies make a vital contribution to any U.S. strategy, even a primarily maritime one. Without them we would have no hope of defending such vital overseas interests as the industrial economies of Western Europe and Japan or their primary source of energy in the Persian Gulf. Without the overseas basing they provide, no viable U.S. force-projection strategy would be feasible. It is often remarked how many overseas bases we have given up. In fact these are mostly marginal compared to those we still occupy.[16]

Our need for allies is greater than ever today, at a time when relative U.S. economic power is declining, Soviet military capabilities are outstripping ours, and nuclear stalemate reduces the credibility of extended nuclear deterrence (see Chapter 2). Indeed, *the single greatest remaining U. S. strategic advantage over the U.S.S.R. is that we are blessed with many rich allies, while the Soviets have only a handful of poor ones.* Most of theirs are a strain on the Soviet exchequer whereas most of ours pay their own way. They also fear their own forced allies, while we fear *for* ours.

Given the traditional reluctance of democracies to fund their own defense adequately in peacetime, there is no realistic way for us and our allies to counter the Soviet military buildup except through a collective effort. It has enabled us to fashion a mutual defense at far less cost than if each nation had to defend itself alone. This collective effort becomes even more imperative when the advent of nuclear stalemate requires greater reliance on much more expensive conventional forces. Hence any realistic strategy and posture for the 1980s and beyond must be based squarely on a collective effort.

Yet, although the Western powers long since definitively adopted a coalition defense policy and strategy, in practice they have never reaped more than a small proportion of the potential for more rational burden-sharing. The United States is as much at fault as most other allies in this respect, but at least it has taken defense more seriously than most of its allies and has shown more will to bear the costs. Thus inadequate allied burden-sharing and inefficient use of constrained coalition resources provide other major reasons why the Western alliance needs to modify its strategy and posture to bring them more into line with what it is collectively willing to spend.

NOTES

1. For example, the first serious French-British staff talks to concert military arrangements for coping with Hitler did not begin until March 29, 1939, only five months before war came. Brian Bond, *British Military Policy between the Two World Wars* (Oxford: Clarendon Press, 1980), pp. 312/313. As another example, Williams describes how U.S. planners in 1916 incredibly assumed that the United States "should prepare to act alone in defending against a German attack." T. Harry Williams, *History of American Wars* (New York: Knopf, 1981), p. 380.

2. Dean Acheson, *Present at the Creation* (New York: Macmillan, 1969), pp. 307/309.

3. Dean Acheson, "Balanced Collective Forces for Europe," in *Present at the Creation*, Chap. 43, p. 398.

4. Ironically, it was France which took the lead in 1950 in "pressing for a united command." Ibid., p. 329.

5. For the best available analysis of various burden-sharingmeasurements see *DoD Reports on Allied Contributions for the Common Defense*, March 1981, March 1982, and March 1983. They analyze NATO and Japanese defense outlays according to standard NATO accounting, which counts all defense spending whether it is in fact allocated to NATO purposes or not. Figures cited are from *DoD Report on Allied Contributions for the Common Defense*, March 1982, p. 40.

6. Ibid., pp. 42/43.

7. For a dissenting view that U.S. pressures created greater political divisiveness than the increased spending was worth, see David Greenwood, "NATO's Three Percent Solution," *Survival*, November/December 1981, pp. 252/260.

8. *DoD Reports,* March 1983, p. 50.

9. Peter Foot, "Problems of Equity in Alliance Arrangements," *Aberdeen Studies in Defense Economics*, No. 23, Summer 1982, p. 1.

10. See General Bernard Rogers, "The Atlantic Alliance: Prescriptions for a Difficult Decade," *Foreign Affairs,* Summer 1982, p. 1148.

11. *DoD Report*, March 1982, p. 11.

12. Ibid., pp. 2/3, 83/85.

13. Foot, "Problems of Equity," p. 41. The reader will also note that I have avoided any concept of "fair" or "equitable" burden-sharing, simply

because there is no adequate formula for measuring it. As this chapter suggests, there are as many intangibles as tangibles; the former defy quantification. The *DoD Report*, March 1981, attempted to construct an ingenious multifaceted "prosperity index" for the purpose, but this has satisfied no one.

14.   *Sunday Times* of London estimates that Tornado costs have risen more than 800 percent in the thirteen years of its development. See "Defense: The Impossible Arithmetic," September 11, 1983, p. 17.

15.   I first advanced the "family of weapons" concept in Annex C of the so-called Komer Report to Secretary Brown in spring 1977, but it was Perry who sparked all the extensive effort to make it work in practice.

16.   From a strategic viewpoint about the only crucially important base we have given up is Dhahran in Saudi Arabia. It proved quite feasible to find substitutes for the rest — e.g., Wheelus in Libya. Moreover, most bases given up were primarily Strategic Air Command bomber or recovery bases (which became less essential as we developed intercontinental bombers and then missiles), or intelligence installations which were superseded by more advanced technical collection means.

CHAPTER 4

What Strategy and Posture?

The factors already discussed, above all the advent of nuclear stalemate, have led to a new ferment about nonnuclear strategy. After long neglect it has again become fashionable. Richard Betts notes how "American conventional strategy and force structure are now subject to more scrutiny than at any time since the 1960s."[1] Huntington finds American strategic thinking, still largely dominated by ideas generated "in the quite different circumstances" of the 1950s and 1960s, of dubious relevance today. He asserts that "American security in the 1980s requires not only a reconstitution of military strength but also a reformulation of military strategy."[2] Sir James Cable, writing of *Britain's Naval Future*, makes a similar plea.[3] Some of the "military reform" group have also focused on this need, especially Congressman Newt Gingrich, Jeffrey Record, and Edward Luttwak. Nor have professional military voices been silent. Admiral Stansfield Turner finds the U.S. military establishment at a historic turning point. It can either "continue with the same strategy that has dominated its thinking, training, and deployment for the last 32 years" or take into account that the world has changed by "revising its strategy through placing more emphasis on the flexibility needed to move forces wherever the United States may require them."[4] Dunn and Staudenmeier call it "the most fundamental debate over American strategy and force posture in more than 30 years."[5]

Perhaps the greatest stimulus has come from nuclear arms control advocates like the so-called American gang of four (Robert Mc-

Namara, McGeorge Bundy, Gerard Smith, and George Kennan), whose advocacy of "no first use" of nuclear weapons is buttressed by their call for more stress on nonnuclear responses.[6] The Reagan administration itself has emphasized the need for a primarily conventional buildup and an accompanying new strategy. Prominent Democratic candidates for president have done the same. Former Vice President Mondale seeks more stress on conventional deterrence, even if more costly than nuclear.[7] Senator John Glenn faults the Reagan administration for allowing force structure and military hardware to dictate strategy rather than formulating a coherent strategy from which force structure and hardware would follow. He too calls for more emphasis on conventional forces. Barry Carter, in a thoughtful essay, makes a similar case.[8]

One argument that is overdrawn, however, is that we need more "strategism" and less "managerialism."[9] We need both. Amen that we need to rethink our strategy, but it is fruitless to decouple this from the need to generate the capabilities to execute it efficiently within likely political and economic constraints. Betts makes a powerful case that strategism and managerialism are interdependent, hence that U.S. stress on the latter is not just "an aberration of American style."[10] In addressing the mismatch between our strategy and our resources, this book too emphasizes not only ways to modify conventional strategy but also ways to generate more efficient outputs out of endemically constrained inputs to enable us to execute it.

Institutional Incapacity of the JCS

Unsurprisingly, however, the institution to which the nation would logically turn as the chief source of such strategic rethinking — the JCS — has been notably unable to tackle the problem. It is important to understand why this is so. It is not because the members of the JCS and Joint Staff are incompetent. Quite the contrary. It is because the JCS as an institution is systemically incapable of dealing with such issues. Because of the way it is constructed, the JCS system is the prisoner of the individual services that comprise it. The rule of unanimity under which the JCS normally operate permits a single service veto, which means in turn that JCS advice almost invariably reflects the lowest common denominator of what the services can jointly agree. The only member no longer beholden primarily to his own service is the chairman, who has little independent authority.[11] The Joint Staff does not even work for him. As a result, the president and his civilian defense leadership cannot rely on getting the unified military advice to which they are entitled — and which they so badly need.[12] This is why JCS reform is so imperative if the nation desires a better military input to strategic decision making and unified military advice on what choices to make between competing service claims in a context of constrained resources.

At present, when the JCS rightly complain about the "mismatch between strategy and resources," their solution is either to duck resource allocation issues or (amounting to the same thing) to call for enough added resources to execute the same old multifront, multiservice strategy which results from splicing together individual service desires. As a result, we have more like *four service strategies* or, more accurately in some cases, strategic doctrines. The navy institutionally focuses on command of the seas, the marine corps jealously fights for amphibious assault, the air force stresses victory through air power (independent if possible), and the army (spread thin over a variety of commitments with a low priority for modernization) looks to mobilizing large forces for sustained overseas campaigns. General E. C. Meyer, when a sitting member of the JCS, found these differences "so wide as to question whether we are pursuing any strategy at all."[13]

The inevitable result is a pyramiding of force requirements when the JCS cobble them together in an annual joint document that contains all the service "wish lists" for global conventional war with the U.S.S.R. Reportedly, the JCS now call for nine more carrier battle groups, fourteen more tactical air wings, and nine more army divisions as needed to carry out the Reagan administration's expansive strategic guidance. This is why a 1982 DoD study that was leaked concluded that a staggering $750 billion more than even the Reagan $1.8 trillion FYDP would be needed to meet these requirements for a total of twenty-three carriers, forty air wings, and thirty-three Army divisions.[14]

These huge requirements are also driven by U.S. military propensities for *worst-case* thinking, especially that any direct U.S.-Soviet regional clash would almost inevitably lead to worldwide nonnuclear (if not nuclear) war. This concept has been a feature of JCS thinking since about the end of World War II. It should not be blamed on the Reagan administration or on notions of "horizontal escalation" on our part. For example, the JCS General Purpose Forces Study of 1962, which foresaw eleven separate theaters as requiring a U.S. force contribution, assumed that "all eleven theaters would come under attack more or less simultaneously."[15] Naturally, we must hedge against the possibility of simultaneous multifront war, but one major strategic issue that needs to be reviewed is whether there are ways to avoid suh a massive diffusion of constrained resources (see Chapter 8).

Service strategic thinking tends to be driven more by competition over how to divide up the constrained defense pie than by clear-headed strategic analysis. The latter is also inhibited by the failure of our military educational system to focus very systematically on strategy, for example through strategic war games or the study of military history.[16] As Liddell Hart put it, "In all our military training . . . we invert the true order of thought — considering techniques first, tactics second, and strategy last."[17] Only recently have our senior military colleges been paying more attention to this subject. The Naval War College has been an exception, although it naturally focuses mostly on maritime strategy. Naval war games have been played since the early 1900s, and Admiral Nimitz is credited with saying that every move in World War II in the Pacific (even Pearl Harbor) had already been played out at one time or another at Newport.

To a considerable extent our neglect of strategy also has been a cultural and societal matter. Political constraints inherent in our system make it difficult to achieve consensus on a coherent strategy. Betts finds that

> the problem lies in the juncture between the American political system and the ambiguity of conventional military requirements for a superpower in a world of both nuclear risks and changing commitments. Ambiguity fosters diverse notions of deterrence and defensive options while democratic politics makes the dominance of any view ebb and flow. Only if U.S. administrations had the duration and consistency of the Soviet Politburo, or if Americans really saw their survival as being tenuous, could there be much more persistent congruence between U.S. strategy and force structure, and thus more room for subtle tuning of doctrine and tactics to strategic guidance.[18]

Although the first condition is hardly likely, the second is. It is precisely the fact that our survival is becoming more tenuous that argues for us to focus on such tuning. The question is whether we will show enough foresight to do so before a major crisis, or will end up being compelled to do so only when it may be too late — and certainly will be far more costly. As Sir James Cable reminds us, "those who refuse to consider alternative courses of action when they can — in time of peace — will be denied the chance when they want to" in a crisis.[19]

The Necessity For Choice

However, revamping U.S. strategy is an enormously complex matter. Aside from daunting issues of linkage between conventional and nuclear responses, the sheer global extent of U.S. interests greatly complicates the task. Few nations have had so wide a range of concerns to deal with — certainly not most of our allies. Our planners face such a range of nonnuclear contingencies that must be deterred, or for which U.S. responses must be designed, that the temptation has been to design general-purpose forces and to emphasize the virtues of flexibility. Given all the constraints on U.S. resources and freedom of action, however, this does not relieve us from the necessity for choice.

Indeed, the essence of real-life strategic decision making is to face up to the hard choices among competing needs in the context of constrained resources. To the extent that more resources are not available, this dictates that we rank our strategic aims in order of priority for resource allocation. Defense "is a question of priorities."[20]

Before addressing the main strategic options open to us, however, it is necessary to mention two strategic concepts which at least theoretically address the central issue. One is whether the United States could regain meaningful strategic nuclear superiority, in which case such great emphasis on costly conventional capabilities would not be so necessary. Against the Soviet Union of today this is quite unlikely, given Soviet ability to match whatever effort we might undertake. Harold Brown, who ought to know about such matters, flatly asserts that "it will not be possible to reacquire strategic and tactical nuclear superiority over the U.S.S.R. Even the attempt to do so could be quite risky and destabilizing." Therefore, he argues against repeating our earlier mistake in the 1950s of concentrating excessively on strategic nuclear capabilities at the expense of conventional.[21]

The other nonstarter would be a frank retreat to neoisolationism, contracting our strategic perimeter to the Western Hemisphere. The most articulate exponent of this school of thought is Professor Earl Ravenal, who sees America as "a mature imperial . . . power, beset by multiple challenges but unable, or unwilling to generate sufficient resources for the defense of its extensive perimeter." His candid answer to this dilemma is to abandon containment and restrict our commitments to what we can readily defend at realistic cost.[22]

The trouble with this concept is that it ignores the long-term balance-of-power consequences of giving up on defense of Europe, Japan, and Persian Gulf oil. It would force our allies to accommodate to the U.S.S.R. Ravenal's approach also ignores our growing interdependence with the world economy in both trade and need for overseas materials. In short, it is a recipe for ultimate defeat or decline.

The Two Main Contending Schools

Instead, service competition for constrained resources is driving a different sort of strategic debate — an old issue in contemporary guise. Two main schools of thought have been reemerging in the conventional arena. One school advocates allocating the bulk of our resources to carrying out a maritime-supremacy strategy aimed not just at command of the seas but also at using them as the major means of offensive force projection against the Soviet Union. Tacitly acknowledging Soviet military predominance on the Eurasian landmass, it stresses U.S. exploitation of the medium we can most readily dominate — the sea (see Chapter 6).

The other school believes that only a continued major U.S. commitment to help defend Western Europe, Japan, and Persian Gulf oil, which implies a more balanced emphasis on land and air as well as naval forces, will suffice to maintain a satisfactory global balance of power. Ergo, it calls for trying harder to generate a credible conventional defense of such vital interests as Western Europe, Northeast Asia, and Persian Gulf oil, primarily via greater and more rational coalition burden-sharing (see Chapter 8).[23]

Both schools accept the premise that preserving credible deterrence now requires the United States and its allies to be prepared to fight a conventional war with the U.S.S.R. — not just achieve a brief conventional pause to underline the cost of crossing the nuclear threshold. Although neither school proposes to abandon nuclear deterrence or the maintenance of adequate capabilities for this purpose, both recognize that, given nuclear stalemate, such capabilities more reliably deter the other side from nuclear escalation than they deter conventional conflict.

To deter or cope with conventional conflict, both schools continue to regard a "forward strategy" based on force-projection capabilities as the best means of damage-limiting as well as of achieving our objectives. Put crudely, both reflect the premise that a big nuclear war is all too likely to engulf our own territory, whereas a big conventional war would necessarily be fought primarily overseas.

Where they differ sharply is over what kind of forward strategy and force-projection capabilities will best serve U.S. interests. Although both schools accept that maritime superiority is indispensable, they strongly disagree over what this means and what kind of navy is essential for the purpose — a sea-control navy or one designed primarily for offensive force projection against the U.S.S.R. (see Chapter 6).

Both these schools represent legitimate responses to the mismatch between our strategy and our resources, in that each accepts that the United States is unlikely to fund both.[24] Unsurprisingly, one reflects the dominant view in the U.S. Navy and the other is often articulated by the U.S. Army. The remainder of this book will deal chiefly with these two schools of strategic thought and various related issues that affect U.S. ability to pursue an optimum nonnuclear strategy in the turbulent period ahead.

NOTES

1. R. K. Betts, "Conventional Strategy: New Critics, Old Choices," in *International Security*, Spring 1983, p. 140.

2. S. P. Huntington (ed.), *The Strategic Imperative*, (Cambridge, Mass.: Ballinger 1982). See Chap. 1, "The Renewal of Strategy," pp. 1/50, esp. pp. 2/3.

3. James Cable, *Britain's Naval Future*, (London: Naval Institute Press, 1983), pp. xv/xvi.

4. Stansfield Turner and George Thibault, "Preparing For The Unexpected: The Need For a New Military Strategy," *Foreign Affairs*, Fall 1982, pp. 122/135.

5. K. Dunn and Colonel W. O. Staudenmeier, "Strategy For Survival," *Foreign Affairs*, Fall 1983, pp. 22/41.

6. McNamara in particular has been strongly criticized for not spelling out convincingly just how the United States and its allies should improve their conventional forces. See Michael R. Gordon's critique ("McNamara's Line," *National Journal*, September 24, 1983, p. 1956) of McNamara's latest article on "The Military Role of Nuclear Weapons," *Foreign Affairs* , Fall 1983, pp. 59/80.

7. Speech before American Newspaper Publishers Association excerpted in *New York Times*, April 27, 1983, p. 16.

8. John Glenn, Barry Carter, and Robert Komer, *Rethinking Defense and Conventional Forces*, (Washington, D.C.: Center for National Policy 1983), pp. 10/13.

9. For a stimulating argument against managerialism, doubtless partly with tongue in cheek, see Edward Luttwak, "Why We Need More 'Waste, Fraud, and Mismanagement' in the Pentagon," *Commentary*, February 1982, pp. 17/30.

10. Betts, "Conventional Strategy," pp. 150/155.

11. This is not to say that some chairmen do not have considerable personal influence. I would argue for example, that General David Jones had more than any chairman since Admiral Arthur Radford. But contrast the chairman's feeble institutional clout with the British system, where the Chief of Defense Staff is the chief military advisor to the Minister of Defense and Cabinet, and has a large central staff which reports directly to him.

12. See *Hearings Before the Investigations Subcommittee of House Armed Services Committee* on "Reorganization Proposals for the Joint Chiefs of Staff" (HASC 97-47 1982), where a parade of experienced witnesses, the author included, have criticized the JCS system.

13.   Cited in John M. Collins, *U.S. Defense Planning: A Critique* (Boulder, Colo.: Westview Press, 1982), p. 157.

14.   George Wilson, "U.S. Defense Paper Cites Gap Between Rhetoric, Intentions," *Washington Post* , May 27, 1982, p. A-4.

15.   W. W. Kaufmann, *Planning Conventional Forces 1950-1980* (Washington, D.C.: Brookings Institution, 1982), pp. 5/6.

16.   Collins, *U.S. Defense Planning*, Chapter 12. He harshly concludes (p. 144) that "no school of strategy in the U.S. prepared senior military officers . . . to perform professionally in that essential field." Cable, in *Britain's Naval Future*, pp. 6/7, laments a similar lack of focus on strategy in British service schools.

17.   B. H. Liddell Hart, *Thoughts on War*, (London: Faber and Faber, 1944), p. 129.

18.   Betts, "Conventional Strategy," pp. 149/153.

19.   Cable, *Britain's Naval Future*, p. xv. 20. Ibid., p. 64.

21.   Harold Brown, *Thinking about National Security* (Boulder, Colo.. Westview Press, 1983), pp. 275/276.

22.   See, for example, Earl C. Ravenal, "The Case for a Withdrawal of Our Forces," *New York Times Magazine*, March 6, 1983, pp. 58/61.

23.   Dunn and Staudenmeier, in "Strategy for Survival," pp. 21/22, also see these as the two main contending schools, although they define their components very differently than I do here.

24.   See Jeffrey Record and Robert J. Hanks, *U S Strategy At The Crossroads.* (Washington, D.C.: Institute for Foreign Policy Analysis, July 1982), for a straightforward plea for a sea-based force-projection strategy "even at the expense of significant reductions in NATO-committed Army forces."

CHAPTER 5

Maritime Strategy or Continental Commitment in Britain and Japan

This tension between "maritime" and "continental" schools of strategy has long historical antecedents. Those Americans who refuse to take it as more than a theoretical construct are simply demonstrating our well-known lack of much sense of history.[1] Similarly, those who persist in seeing it as merely a facade for the perennial interservice competition for resources simply illustrate my contention that we Americans don't think much about strategy either. A brief look at history is instructive. It shows that the conflict between these two strategic approaches is about as old as the history of sea power itself.[2]

> There have been over history two kinds of conquerors, two kinds of nomads: the horsemen and the seafarers . . . the vicissitudes of power are controlled by the struggle of land and sea, the victory going to one or the other in turn, depending on whether the continental or the maritime power possesses more resources and whether technology favors one or the other.[3]

Thucydides' *History of the Peloponnesian War* cites perhaps the first historical example of how overconfidence in superior sea power and its consequent misuse in distant adventure led to the downfall of Athenian supremacy. All the great empires of the ancient and medieval worlds were predominantly land powers, although in the Punic Wars Rome was long checked by Carthage until it also achieved a naval dominance permitting the projection of its land power overseas.

But what is generally regarded as the great age of sea power occurred between 1500 and 1900. "Sea power exerted its greatest influence between the early sixteenth and later nineteenth centuries, that is between the creation of the oceanic sailing ship on the one hand and the industrialization of continental land masses on the other."[4] Mahan drew his lessons mostly from this age of sail, using Great Britain as his classic case. In his brilliant analysis of *The Rise and Fall of British Naval Mastery*, Paul Kennedy assails Mahan's attempt to draw universal axioms about the "superiority of the sea over the land" from only this one historical period, also one in which only a relatively small group of states — Portugal, Spain, Holland, France, and England — were involved.[5] Land warfare remained the dominant form of combat even in the age of sail.

Navalists versus Continentalists in Whitehall

The competition between sea and land power claimants on the limited defense spending available was most prominent in Great Britain. Kennedy puts it succinctly as a

> centuries-long debate between what have been termed the "maritime" and the "continental" schools of strategy: that is, between those who have maintained that Britain should concentrate her energies upon her navy, her colonies and her overseas trade, remaining aloof from Europe in peacetime and only carrying out peripheral raids against the enemy and offering subsidies to allies in wartime; and those who have held that under certain circumstances a continental military commitment was necessary, since their country's own security was inextricably bound up with the fate of the European balance of power, and that an isolationist policy would endanger Britain too in the long run. This debate, about the balance that should be reached between land power and sea power, between Europe and the wider world, between the army and the navy, was always a contentious one, for it had personal, emotional and domestic political aspects and repercussions which took it outside the realm of pure strategy. My own position, no doubt influenced by events in the present century, is one of support for those who, like Elizabeth I, William III, Marlborough, Chatham, Grey and others, argued the need for the British people to balance their natural wish for a "maritime" way of life and strategy with a watchful concern for Europe and a determination to ensure that developments on that continent did not deleteriously affect their country's interests.[6]

The Royal Navy had its real inception in the early sixteenth century under Henry VIII. Under Henry and then Queen Elizabeth was also born "the 'navalist' or 'Blue Water' school of strategic thought and its conflict with the continental school."[7] Indeed, King Henry and his adviser Cardinal Wolsey also recognized the need to preserve a favorable balance of power "which should prevent any power from having a hegemony on the continent or controlling the Channel coasts."[8]

The eighteenth Century Anglo-French wars revived the argument between navalists and continentalists, the

> central strategic dispute which had so divided the Elizabethans and which only the peculiar nature of the intervening Anglo-Dutch conflicts had temporarily obscured; whether Britain's policy in a European war was to become militarily involved to a large degree upon the mainland in support of allies, or whether it was better to adopt a "maritime" or "Blue Water" strategy of colonial conquest, commercial pressure and naval victories instead.[9]

Even the continental successes of Marlborough (that great exponent of coalition war) in the War of the Spanish Succession did not bring home this crucial connection "between Britain's European and naval policy." "Usually the proponents of each school were to scorn the arguments of their rivals — to the detriment of a balanced, harmonious policy."[10]

Throughout the seven Anglo-French wars ending at Waterloo in 1815, except for a brief period at their outset, Britain dominated the seas. But this alone never proved decisive against a primarily continental power like France. On the other hand, the French privateering *guerre de course* made substantial inroads on British trade, an analogy to Germany's submarine campaigns of 1914/1918 and 1939/1945. Hence statesmen like Pitt the elder pursued a mixed continental and maritime strategy, in which large British subsidies of continental allies like Frederick the Great were often supplemented by British armies on land.

Britain's wars against Revolutionary and then Napoleonic France led to the zenith of its naval supremacy. British domination of the seas permitted a remarkable expansion of Britain's empire. However, "years of peripheral and overseas operations revealed that Napoleon could only really be defeated on land," and "the ancient dispute between sea power and land power reemerged in heightened form."[11] Again the defeat of a great continental land power demanded not only naval mastery but also the building and rebuilding of continental alliances backed not only by unprecedented British subsidies but eventually by sizable British land forces to join her allies. Moreover, only Napoleon's disastrous attempt to invade Russia made ultimate victory possible. Until the Peninsular Campaign of 1809/1813, few of Britain's many amphibious raids really affected the overall European balance of power. To cite Kennedy again, "Only the patriotic bias of British naval historians makes it necessary to point out the truism that a war for the military domination of Europe had to be fought, logically enough, inside Europe and by armies. To defeat Napoleon, 'maritime' methods had to be supplemented by 'continental' ones."[12] Throughout the seven Anglo-French wars, outlays on the British Army greatly exceeded those on the Royal Navy. In sum, although command of the sea proved enormously valuable in containing and ultimately defeating Napoleon, "the conflict with France also confirmed the limitations of sea power, the necessity of watching carefully the European equilibrium, the desirability of having strong military allies in wartime, the need to blend a 'maritime' strategy with a 'continental' one."[13]

Nor should it be forgotten that during the sixteenth to nineteenth centuries when navalism was so predominant, Britain in effect "bought" mercenary armies to help maintain a balance of power. Arguably these large subsidies to continental allies were national security outlays. The modern equivalent, for Britain and to a far larger extent the United States, is military aid plus a good portion of so-called economic aid (see Chapter 9).

The supremacy of the Royal Navy throughout the nineteenth century fortified the contentions of the scholars of naval strategy who emerged in the later years of that century. The most prominent was Mahan, who enjoyed a greater vogue in England than in his own country. His ideas "centered on the belief that sea power had been more influential than land power in the past and would always continue to be so."[14] Mahan's ideas were far more widely known than those of Sir Halford Mackinder, who wrote of the coming dominance of large landmasses and populations — particularly the Eurasian "heartland."[15] Mackinder's apocalyptic views did little to shake British complacency, which was

> reinforced by the intellectual superiority of the advocates of the "Blue Water" or navalist school over their "Brick and Mortar" or army rivals, a victory which led not only to the reversal of the mid-nineteenth century policy of raising a militia and building fortifications against possible invasion but also to an almost absolute belief in the effectiveness of sea power.[16]

Kennedy sees the root cause of the subsequent gradual decline of Britain's naval power in the decline of her comparative economic strength, "since other nations with greater resources and manpower were rapidly overhauling her previous industrial lead."[17] This problem was accentuated by the rising cost of modern armaments and, as Mackinder contended, the waning of sea power in relation to land power with the advent of railways, mines, torpedoes, and later submarines and aircraft.

The immediate cause, however, was the rise of another potentially dominant power in Europe. After defeating France in 1870, imperial Germany soon turned to building a modern high-seas fleet that challenged British naval supremacy in home waters. It is often forgotten how quick and far-reaching was Britain's shrewd adjustment to this reality. The "vast reorientation of British naval policy" included giving up the so-called two-power naval standard, abandoning splendid isolation for a policy of alliances (the first was the Anglo-Japanese alliance of 1902), and redeploying the fleet largely to home waters to counter Germany, while preparing again to contribute a sizable army to the defense of France and the Low Countries. By 1912 the Royal Navy had even brought home most of its Mediterranean Fleet, following First Lord of the Admiralty Churchill's principle that "if we win the big battle in the decisive theater we can put everything straight afterwards."[18] What made this possible was the Anglo-French entente, under which the French navy assumed primary responsibility for the Mediterranean (an early example of specialization of mission within a coalition). In short, Britain felt compelled to cut its naval coat to fit the strategic cloth.

The rise of imperial Germany also revived the debate between the maritime and continental schools of strategy. Britain again had to grapple with whether to send an army to the continent, this time to help France maintain a balance of power vis-a-vis Germany. Only by reinforcing the French army on the continent could Britain preserve an adequate balance of power, as was proved by World War I. However, all this was strongly opposed by the isolationist admiralty, especially First Sea Lord Fisher.

World War I demonstrated, in the words of Admiral Richmond, that sea power and land power were interdependent and that a purely peripheral maritime strategy alone would have permitted German conquest of Europe.[19] Nor was sea power decisive. The British Grand Fleet did contain the German High Seas Fleet, and distant naval blockade did contribute to allied victory, but it probably did less to weaken the Central Powers than German U-boat campaigns did to weaken the allies. Gallipoli — the only truly amphibious major operation of the war — ended in disaster. Most of all, the horrendous costs of World War I spelled the end of Britain's mastery of the seas.[20]

Postwar U.K. governments were quite unwilling to fund a dominant navy. Its budget was repeatedly and severely pruned under the notorious Ten-Year Rule, whereby each year from 1919 to 1932 the British Cabinet told the Chiefs of Staff that they need not assume any major war for the next decade (hence need not prepare for one). All that kept Britain in the front rank of naval powers was the 5-5-3 tonnage ratio between America, Britain, and Japan imposed by the Washington naval agreement of 1923.

Another serious threat to both navy and army budgets was the favored competitive position of the newly independent Royal Air Force (RAF), whose independence — as well as its absorption of the fleet air arm — was naturally opposed by the Admiralty and War Office. But successive British cabinets accepted the thesis that Britain's greatest vulnerability was from the air. By 1938 the RAF budget exceeded that of the army or the navy.

The army fared even worse than the navy. Britain's staggering manpower losses on the Western Front led to a revulsion against any continental commitment. It was now "bitterly condemned as an unwise — worse still, an unnecessary — departure from an allegedly traditional maritime strategy in which small military expeditionary forces played a valuable but subsidiary role." "Never again" became public preference as well as official policy, and the army was reduced to little more than an imperial police force.[21]

Even after the rise of Hitler, British governments in the middle and later 1930s allowed wishful thinking to blind them to strategic reality.[22] Britain's overall unreadiness made even the chiefs of staff "appeasers" on strategic grounds. They made clear to the cabinet that Britain would be spread much too thin to fight a three-front war against Germany, Italy, and Japan simultaneously.[23] Admiral Chatfield, when he served as First Sea Lord and then Minister for Defense Coordination, emphasized that "Britain must reduce the number of her potential enemies," a lesson relevant to U.S. strategic planning as we confront possible multifront conflict with the U.S.S.R.[24] The obvious analogy is to our current three-front problem of defending Europe, Japan/Korea, and Persian Gulf oil.

World War II, in which strategic realities dictated that an unready Britain nonetheless again send major forces to the continent, further hastened the Royal Navy's decline. It performed exceptionally, much better than in World War I. By 1942, however, the focus of naval war shifted to the U.S. Navy's brilliant island march across the Pacific after the initial disaster at Pearl Harbor compelled it to shift from battleships to carriers as its *arm blanche*. World War II also saw the zenith of amphibious assault — made feasible by command of the sea and air superiority both in the Channel/Mediterranean and in the Pacific. But again the horrendous economic costs of the war hastened Britain's decline and made it impossible for the Royal Navy to regain first-class status in the age of superpowers. The maturing of air power, which so increased the vulnerability of surface ships, was a further factor.[25]

British defense policy after World War II was far more realistic than after World War I. The hard lessons had been learned. Splendid isolation based on British maritime strength was no longer feasible. Reluctantly, haltingly, Britain turned toward Europe — first strategically, then economically as well. The perceived Soviet threat to the European balance of power dictated not only a major peacetime commitment of forces in Germany, but also a policy of collective security based on firm postwar alliances. By joining the Western European Union and then NATO, Britain opted definitively for a continental commitment. As Michael Howard puts it:

> basically our security remains involved with that of our continental neighbors: for the dominance of the European landmass by an alien and hostile power would make almost impossible the maintenance of our national independence, to say nothing of our capacity to maintain a defensive system to protect any extra-European interests we may retain.[26]

Nevertheless, all three British services were cut recurrently as Britain's economic strength declined further. "Those long term trends detected earlier — the relative weakening of the British economy, the disintegration of the European empires, the decline of sea power in its classical form vis-a-vis land and air power, the growing public demand for increased domestic rather than foreign expenditure — now combined finally to overwhelm a British naval mastery that had long been in question."[27] It disappeared as part of a long recessional, the retreat of Britain from a worldwide empire to mostly regional power status. "The price of Admiralty was too high."[28] Washington had to assume London's earlier strategic responsibilities in the wider world.

"Equally natural was the Admiralty's fight against this line of reasoning, and the implications for the priorities in the country's defence budget. The debates of 1906/1914 and 1920/1939 were being fought out again, but this time the historical precedents could merge with the logic of the existing power balance to overwhelm the Navy's appeal to 'the British way of war.' "[29] The Royal Navy's long rear-guard action naturally extended to opposing creation of a new Ministry of Defense, under which all three services would be integrated. The Admiralty saw a fundamental challenge to its autonomy in the postwar series of decisions (especially those of 1946, 1963, and 1967) to strengthen the Ministry of Defence.[30]

Recently the brief Falkland Islands conflict has rekindled the debate, though even that ardent navalist James Cable dubs it "an exceptional incident which can hardly be used to support the case for a maritime strategy and a much stronger Royal Navy." It was an "aberration . . . from the primary task of defending the British Isles against the various threats which Soviet hostility could pose."[31] After all, how many other colonial remnants does Britain still possess? Nevertheless, several critics have used the Falklands as their justification for calling for an expanded fleet, even if other needs must suffer. Cable, on the other hand, wisely examines general strategy as the setting for naval strategy and his basis for assessing the future utility of the Royal Navy. In *Britain's Naval Future*, he emphasizes the Royal Navy's need to choose intelligently among strategic options when resources are so constrained.[32] This book takes the same approach to assessing U.S. needs.

The Case of Japan

Similar tensions between maritime and continental schools of strategy were also characteristic of that other offshore island nation across the world — as Japan emerged in the nineteenth century from its self-imposed isolation. Here the army was usually dominant. Hence the Japanese pursued primarily a continental strategy aimed at expanding Japan's hold on the adjacent mainland of Asia. The navy's strategy was defensive — to protect the homeland and control the sea-lanes to support the army on the continent.[33] A competition for constrained resources ensued.

Japan's concern over the Chinese and the Russian influence in nearby Korea helped lead to the Sino-Japanese War of 1894/1895 and then the 1904/1905 Russo-Japanese War. "But Admiral Yamamoto, naval minister and real founding father of the Imperial Navy, opposed Japanese involvement in Korean affairs," arguing that so long as Japan controlled the sea, its home territory was safe. If Japan could not control the sea, however, any continental involvement was doomed. Shunji Taoka calls this one of the first statements of a blue-water strategy for Japan.[34] But the Japanese army argued that the chief threat lay from Russia to the west, and wanted to plan against three enemies — Russia, the United States and France, in that order.

The navy saw a single dominant enemy — the United States. "Coordination between the army and navy had been poor in the Russo-Japanese War, and in 1906 the army proposed that Japan adopt an integrated defense policy . . . with a general agreement between the two services on a common strategy." The navy was opposed, even refusing to consult with the army on the conduct of land/sea operations.[35]

As in Britain, the maritime and continentalist schools tended to go their separate ways. The United States became the principal foe against which the navy planned, whereas the army was more concerned with the Russian threat and expansion into China. In 1907 an Imperial Defense Policy decreed Russia to be the chief potential foe, thus prolonging the dominance of the army and the continental school.[36] The navy agreed to this only as a matter of court formality, and continued to plan primarily for conflict with the United States right up to World War II. "The two seldom argued openly with each other but simply took different courses."[37] The army was dominant in the occupation of Manchuria in 1931, as well as the later expansion into China. Japan's admirals held quite another view — that the main threat to the home islands and the sphere of influence in China was from U.S. sea power to the east and south. Hence the navy favored an "advance south" strategy of holding maritime barriers anchored to the Pacific islands and expanding toward the oil and other resources of Southeast Asia. Still, most of the Pacific islands were left unfortified even by 1941, as Admiral Inoue found to his surprise when he was nominated as Commander-in-Chief (CINC) of the 4th Fleet stationed there.[38] In World War II the Imperial Navy operated primarily as a "sea denial" force, "oriented to denying control of the Western Pacific to the U.S. fleet in order to allow the army to win the main war on the continent."[39]

After World War II Japan adopted a U.S.-style Joint Chiefs of Staff system to achieve better coordination among the services. Although the Maritime Self-Defense Force did manage to retain a substantial degree of autonomy in Japan's postwar self-defense arrangements, it has had great difficulty in acquiring respectable capabilities for protecting Japan's sea-lanes, even out to 1,000 miles. In fact, all three services have been unable to date to provide even for adequate defense of the home islands. But even today the Ground Self-Defense Force tends to "look north" and see the main threat as being Soviet attack on Holkaido, while the navy tends to "look south" at the main threat as being Soviet attack on Japan's commerce routes, especially for vital oil imports from the Persian Gulf.[40]

To conclude, even though America's continental size, position, and resources make it unique compared to offshore islands like Britain and Japan, the many disagreements between their armies, navies, and later air forces over strategy and resource allocation offer lessons for the United States as well. Compromise in the interest of a balanced strategy and force posture came hard to either camp. The admirals and their partisans were probably more parochial than the generals, at least in the United Kingdom. However, the purpose of this chapter is not to pass judgment but rather to demonstrate that the tension between maritime and continental schools of strategy is a long-standing historical phenomenon, not just an artificial construct. Moreover, it is hard to fault Brian Bond's "depressing thought that many of the intractable problems confronting the makers of British defence policy in the 1930s are still with us today."[41]

NOTES

1. For a good historical review of maritime strategy and its multinational literature, see Geoffrey Till, *Maritime Strategy and the Nuclear Age* (New York: St. Martin's Press, 1982).

2. Osgood sees it as a manifestation of the perennial attempt to cope with the gap between a nation's interests and the power to support them. Robert E. Osgood, "American Grand Strategy: Patterns, Problems, and Prescriptions," *Naval War College Review*, September/October, 1983. p. 9.

3. Raymond Aron, *Peace and War* (Garden City, New York: Doubleday, 1966), p. 157.

4. Paul M. Kennedy, *The Rise and Fall of British Naval Mastery*, (New York: Scribner's, 1976), p. xvi.

5. Ibid., pp. 6/7. This book is Kennedy's "attempt to carry out the first detailed reconsideration of the history of British sea power" since Mahan. He urges keeping in mind that "Mahan and his followers were writing about an era during which commerce and conflict at sea occupied a disproportionately large role in world affairs." Though acknowledging that Mahan himself "never went so far as his more extreme followers," he holds Mahan's works "largely responsible for a whole school of strategic thought" accepting uncritically the idea that "the sea has played *the* leading role in the advancement of civilization" and for "the frequent deprecation of the role of land power by the navalist school of strategy."

6. Ibid., pp. xvi/xvii.

7. Ibid., pp. 26/29. He defends against the complaints of the "navalists" Elizabeth's balanced view of the need for both naval superiority and continental commitment.

8. Henri Pirenne, *The Tides of History* (New York: Doubleday), Vol. II, p. 429.

9. Ibid., p. 75.

10.   Ibid., p. 75.

11.   Ibid., pp. 129, 132.

12.   Ibid., p. 137.

13.   Ibid., p. 147.

14.   Ibid., p. 182.

15.   Halford Mackinder, "The Geographical Pivot of History," *Geographical Journal*, April 1904. Interestingly, Kennedy titles his Chapter 7, "Mahan vs. Mackinder (1858/79)," in *Rise and Fall*, pp. 177/202.

16.   Kennedy, *Rise and Fall*, p. 185.

17.   Ibid., pp. 185/186. Kennedy develops the compelling thesis that Britain's naval growth and decline was bound up with her economic growth and decline.

18.   Ibid., pp. 223/225. James Cable, *Britain's Naval Future* (London: Naval Institute Press, 1983), pp. 15/16.

19.   Sir Herbert Richmond, *National Policy and Naval Strength and Other Essays* (London: Ernest Benn, 1928), p. 77.

20.   Cable, *Britain's Naval Future* , pp. 16-17.

21.   Brian Bond, *British Military Policy between The Two World Wars*, (Oxford: Clarendon Press, 1980), p. 1.

22.   Ibid., p. 338.

23.   Michael Howard, *The Continental Commitment*, (Oxford: Clarendon Press, 1972), pp. 118/120. Kennedy, *Rise and Fall*, pp. 289/293.

24.   Cable, *Britain s Naval Future*, p. 17.

25.   Kennedy, *Rise and Fall*, Chapter 11. He notes that Britain is estimated to have devoted 50/60 percent of its war production to the RAF.

26.   Howard, *Continental Commitment*, pp. 9/10.

27.   Ibid., p. 324.

28.   Brian Schofield, "The Price of Admiralty," *RUSI Journal* , September 1976, p. 89.

29.   Kennedy, *Rise and Fall* , p. 328.

30.   Ibid., p. 329. Cable, *Britain's Naval Future* , pp. xii/xiii.

31.   Cable, *Britain's Naval Future*, pp. xii-xiii.

32.   Ibid., pp. 8-9, and Chapter 7. He concludes that the four strategic missions of current U.K. forces — the nuclear deterrent, air defense of the U.K., the continental commitment, and the Navy — cannot be maintained with foreseeable resources. "The nuclear deterrent is the most vulnerable," but he favors keeping it and air defense. He tends to favor the navy over the continental commitment (pp. 176/188).

33.   C. G. Reynolds, "The Continental Strategy of Imperial Japan," *U.S. Naval Institute Proceedings*, August 1983, pp. 65/71.

34.   Letter from Shunji Taoka to J. E. Auer, November 18, 1983. The

author is greatly indebted to Commander Auer (USN Ret.) and his excellent Japanese sources, especially Taoka, the senior military writer for *Asahi Shimbun* . Their thoughtful comments on Japanese maritime versus continental strategies were indispensable.

35.   J. E. Auer, *The Postwar Rearmament of Japanese Maritime Forces 1945/71*, Praeger, 1973, pp. 17/18. Taoka differs with Auer on this point, contending that army-navy cooperation was "very good" in the Russo-Japanese War, with the army taking Port Arthur in order to help destroy Russia's Pacific fleet before the Czarist Baltic fleet could arrive.

36.   Reynolds, "Continental Strategy," p. 67.

37.   Taoka letter of November 18, 1983.

38.   Letter, Jun Tsunoda to J. E. Auer, November 14, 1983, commenting on author's manuscript.

39.   Reynolds, "Continental Strategy," p. 71. Taoka's letter adds that the army, being uninterested in the south, did not conduct its first map exercise on invading the Philippines until November 1941, a month before the actual invasion.

40.   Taoka letter of November 18, 1983.

41.   Bond, *British Military Policy*, p. 339.

CHAPTER 6

The New Navalism: Emergence of a U.S. Maritime Strategy

In the United States tensions between these two competing schools of strategy did not become a major factor until World War II. More investment usually went to the peacetime navy than to the army, a natural consequence of the fact that the Atlantic and Pacific oceans provided our geographic security against any major power. Indeed, a sizable part of the army's mission up to World War II was to provide coast defense of key ports to enable the fleet to operate flexibly and provide it secure bases.[1] Ad hoc army-navy cooperation occurred successfully in the Civil War but was disgracefully lacking in the 1898 campaign against Santiago, Cuba during the Spanish-American War.[2] Not until the 1903 creation of the first interservice planning body, the Joint Army and Navy Board, was peacetime joint planning even systematically addressed.[3]

Early Strategic Disputes

Four years later the Joint Board became the scene of the first major interservice conflict — over whether the navy's main Pacific fleet base should be at Subic Bay in the Philippines or someplace the army could defend more readily. Then the army and navy clashed over Billy Mitchell's aggressive case that land-based bombers could sink battleships, hence that air power — not sea power — should become our first line of defense.[4] However, serious strategic differences between the services did not really surface until World War II, when the army and navy quarreled repeatedly over the allocation of resources to the Pacific versus European theaters (see Chapter 1). These differences were exacerbated as the postwar United States succeeded to Britain's and Japan's global strategic roles.

Given postwar budget stringency, the issue tended to be framed in terms of competing claims for the constrained resources available in peacetime. As in Britain, it became a three-way battle, especially after the 1947 advent of an independent U.S. Air Force. In the 1950/1951 "great debate" over stationing U.S. troops in Europe, both ex-President Hoover and Senator Robert Taft argued vigorously for primary reliance on U.S. sea and air forces instead.[5] Underpinning this debate were maritime versus continental issues quite reminiscent of the long British debate, plus similar interservice quarrels over roles and missions. The competition between land-based and sea-based air power arose again when U.S. Air Force programming of intercontinental bombers triggered the notorious "revolt of the Admirals." President Truman's proposed defense budget of only $11 billion for FY 1949 "set the stage for a bitter interservice debate about roles, strategy, and finance."[6]

NSC-68, America's first peacetime national strategy, did not resolve these differences. Nor did creation of an overarching Department of Defense in 1947 help much. To this day the U.S. Navy has continually resisted efforts to integrate the defense establishment. Its opposition to creation of an independent air force, to giving the Joint Chiefs of Staff real power, and even to creation of a Department of Defense all reflected fear that naval autonomy would be compromised, similar to the British Admiralty's earlier opposition to such developments in the United Kingdom.[7] In the frustrated view of Henry Stimson, the distinguished Secretary of War in World War II, "The Navy Department . . . frequently seemed to retire from the realm of logic into a dim religious world in which Neptune was God, Mahan his prophet, and the United States Navy the only true church."[8]

Interservice strategic differences continued to revolve around the question of deciding which theaters were to have priority. The army, fixed on defense of NATO Europe, favored "global" planning. It wanted the annual midrange Joint Strategic Objectives Plan (JSOP) prepared by the JCS to have clear regional priorities. The navy tended to favor "parametric" planning, simply to sum up the war plans of the regional Commander-in-Chiefs (CINCs) as the basis for an overall war plan, and to preserve flexibility to adjust to changing circumstances.[9] For the latter reason the navy continually opposed assigning forces in peacetime to nonnaval CINCs (most recently to the Rapid

Deployment Force). An equally important issue has been the long-standing army concern over adequate sea-lane protection to permit army/marine forces to be projected overseas in timely fashion, versus the navy's dominant preference for maritime power projection via carrier battle groups even at the expense of the capabilities of the other services. Almost by definition the army has preferred a sea-control navy. So, for its own reasons, has the U.S. Air Force. Similarly, the army has long sought more navy and air force emphasis on sea lift and airlift as essential to deploying the army overseas in any force-projection role.

Public debate over issues such as these above was muted during the period of primary U.S. reliance on nuclear deterrence. However, as nuclear stalemate compels the United States and its allies to pay greater attention to coping with Soviet conventional superiority in Eurasia as well as with growing Soviet sea power, differing army-navy views over conventional strategic issues are assuming greater prominence.

Strategic Issues in Harold Brown's Pentagon

Although the Carter administration has been accused of being "soft" on defense, it did succeed in reversing the postVietnam decline in U.S. defense spending, achieving 10 percent real growth in four years. Moreover, it also tried to address the strategy side of the strategy/resource mismatch. Under Harold Brown — the most experienced secretary of defense since George Marshall and regarded as the strong man of the Carter cabinet — DoD tended to dominate the evolution of conventional as well as nuclear strategy.[10] Acutely conscious of budget realities and of President Carter's reluctance to increase defense spending as sharply as the security situation by itself would demand, Brown and his team sought to optimize U.S. posture within the budget constraints under which they labored.

For one thing, we were strong advocates of a coalition strategy based on more rational burden-sharing with our allies, especially Europe and Japan. Efforts to promote such burden-sharing were vigorously pursued. The Pentagon was the source of the Carter initiatives to rejuvenate NATO's defense posture (see Chapter 3). Indeed, it used NATO agreement on a goal of 3 percent real growth in defense spending as an effective lever to get increased U.S. defense outlays.

However, the policy focus on rebuilding NATO defenses was not reflected in any great shift in U.S. resource allocations to NATO at the expense of other theaters. Nor did the Pentagon adopt any more of a "NATO first" or "Central Front" strategy than had traditionally been the case since the end of World War II.[11] True, the administration did launch numerous initiatives designed to remedy the long neglect of our European flank during our decade-long entanglement in Southeast Asia (see Chapter 3). But no planning or program document from the NSC, the Pentagon, or the JCS ever called for a predominantly central-front strategy. More important, the traditional division of funds between the services did not significantly alter. Any NATO emphasis was rather on speeding up the first wave of already long-planned NATO reinforcement by prepositioning more equipment in Europe and getting more allied host nation support; the largest single increase in NATO-oriented funding was for such prepositioned equipment.

Indeed, NATO again took a back seat when the Pentagon began in 1980 giving first priority to coping with the power vacuum in the Persian Gulf. The army itself urged shifting program funds from additional prepositioning in Europe to the new RDF buildup, and some of this was done. Several army divisions previously earmarked for contingency deployment to Europe as first priority were instead given first-priority missions in the RDF. The same happened to some air and marine forces. In short, if there was a significant refocusing of U.S. forces and resources, it was toward the Persian Gulf rather than Europe.

For similar reasons Secretary Brown modified the so-called "swing strategy," whereby in event of a NATO/Warsaw Pact war, even if it became global, a large number of Pacific-based forces (especially carrier battle groups) would be shifted to the Atlantic. The navy was always uncomfortable with depriving the Pacific theater of so many forces and finally prevailed on the JCS to propose formal abandonment of a mandatory swing. In 1979 Secretary Brown accepted that swinging would henceforth be only one option, not in order to leave the forces concerned in the Pacific as the navy desired, but to take into account the looming requirement for naval power in the vulnerable Persian Gulf-Indian Ocean region.

Nor did the navy's budget suffer from the Carter administration's alleged Eurocentrism. Contrary to later allegations,

from 1978 to 1981 the U.S. Navy enjoyed higher spending and more purchasing power than for any comparable period since World War II. Not only did navy budget authority rise from $33 billion in 1977 to nearly $53 billion in 1981 — growth averaging nearly 4 percent per year above inflation — but the sea service garnered a greater share of the defense funds added during the period than did any other service. . . . Spending on ship construction, missiles, mines, and other new weaponry grew by almost two-thirds, while funding for new navy aircraft more than doubled.

Moreover, the Carter administration's last FYDP called for more than 5 percent real annual growth in further navy spending for 1982/1986.[12] Nonetheless, the Brown Pentagon did seek to resolve the issue of how many carrier battle groups (CVBGs). This argument was not over whether command of the sea at times and places of our own choosing was essential; obviously it was. Rather, it revolved around whether roughly simultaneous carrier operations in three or four widely separated theaters were essential at the outset of a major U.S.-Soviet conventional conflict, or whether hitting the enemy first in one place and then shifting carriers from one sea to another, sequentially capitalizing on the innate flexibility of sea power, would suffice. Given constrained resources and the need for a balanced multiservice posture, Secretary Brown's *Defense Guidance* called for sequentialism rather than simultaneity as the optimum, which justified a force of twelve modern carrier battle groups.[13] And the Carter administration did cancel one planned nuclear-powered carrier in 1978.

The Reagan Strategy and Posture

Few of the active or latent differences over U.S. conventional strategy have yet been resolved by the Weinberger Pentagon either. Indeed, far from resolving them, it appears to have accentuated them by espousing an ambitious all-inclusive global strategy. Emphasizing the break with "obsolete strategic concepts," Secretary of Defense Weinberger called for "a break with some past thinking . . . to develop new policy and concepts."[14] On top of an ambitious strategic nuclear modernization program, he has favored greater emphasis on conventional forces able to "meet the demands of a worldwide war, including concurrent reinforcement of Europe, deployment to Southwest Asia . . . and support in other potential areas of conflict."[15] Further, he has attacked what he calls "the short war fallacy," emphasizing the costly need to improve sustainability for protracted conventional as well as nuclear war.[16]

In short, the secretary of defense has seemingly accepted all the strategic missions called for by the JCS, which in turn reflect an amalgam of parochial service views cemented by the conviction that

any direct clash between the two superpowers would almost automatically become global in nature, wherever it began (for a contrary
view see Chapter 7). Even discounting the rhetoric designed to dif-
ferentiate it from that of previous Republican as well as Democratic
administrations, this strategy is so divorced from real-life budget
constraints as to smack of an unconstrained theoretical exercise.

The Reagan administration deserves full credit for accelerating
the effort to strengthen U.S. defenses, but even if its programmed
buildup were fully funded, this would not give the United States
enough capability to meet our needs on several fronts simultaneously
(except at sea). Nor would it suffice for protracted conventional conflict. Qualified critics have repeatedly made this point.[17] This is why
in 1982 the JCS reportedly called for forces estimated to cost $750
billion beyond the $1.8 trillion programmed for FY 1983/1987 in
order to carry out this strategy — including nine more carrier battle
groups, nine more army divisions, and fourteen more tactical air
wings.[18] Thus, despite its efforts to increase U.S. defense spending
sharply, the current administration has not managed to resolve the
mismatch between our conventional strategy and resources either.

An overall NSC review of national security strategy in February/
May 1982 took a more sober and balanced view of military priorities.
While calling for a strengthened nuclear deterrent, this review found
"conventional" deterrence to be now more important than ever, and
stressed the need for a "coalition strategy" to this end. It also acknowledged that despite the risk of global war, we could not expect
"to successfully engage Soviet forces simultaneously on all fronts."
This meant that "we must procure balanced forces and establish
priorities for sequential operations to ensure that military power would
be applied in the most effective way." The review also found it

> in the interest of the United States to limit the scope of any
> conflict. The capability for counteroffensives on other fronts is
> an essential element of our strategy, but is not a substitute for
> adequate military capability to defend our interests in the area
> in which they are threatened.[19]

This sober statement of the military aspects of our national strategy
(quite reminiscent of that of previous administrations) seems to strike
a somewhat different note from what the Pentagon is actually funding.

We must look, therefore, at *where the Reagan administration allocated the bulk of its conventional force funding* to get a clue to what
strategy actually could be pursued. After all, posture tends to drive
strategy more than the reverse. While the administration clings of-

ficially to an eclectic strategy, and Secretary Weinberger regularly reassures NATO of the U.S. commitment, the greatest effort is being allotted to a naval buildup. As Secretary Weinberger puts it, "the most significant force expansion proposed by the administration centers on the navy, particularly those components of it which have offensive missions."[20]

"The Navy Is Going Wild"[21]

To this end, Secretary of the Navy Lehman has moved smartly to lock in a "600-ship" naval program, including 15 carrier battle groups (CVBGs) and sizably increased amphibious assault capabilities.[22] Secretary Weinberger's decentralization of authority back to the services greatly facilitated his ability to do so. "Front-loading" by getting funds to start these ships included in the new administration's first two defense budgets (FY 1982 and 1983) enabled Lehman (by his own proud admission) to get his shipbuilding program under contract almost regardless of later cuts in proposed DoD spending.[23]

In FY 1983, for example, the navy's budget share rose to over 34 percent, compared to 32.5 percent for the air force and only 23.7 percent for the army, primarily to fund two new carriers in one year. The disparity between the navy and the army is even greater when one compares their *investment* budgets. Here the navy was allotted roughly double what the army received in FY 1983.[24] From a parochial viewpoint Lehman did a remarkable job for the navy, or at least for the dominant "brown shoe" aviation faction in it.

His predictions that his shipbuilding program would not be significantly altered in future budgets have been borne out in the FY 1984 budget authorization, where congressional insistence on allowing only 5 percent real defense budget growth instead of the 10 percent the administration asked for has resulted in the other services' programs being cut back much more than the navy's.[25] Although the other services also received added funding in FY 1982 and 1983, much of their modernization program and planned force-structure increases were deferred to the outyears.

Now it seems questionable whether their programs can be carried out. In the FY 1984 budget, the navy's shipbuilding program has hardly been touched, but army and air force conventional requests have been slashed. Moreover, the Pentagon's Defense Resources Board, in drawing up the FY 1985 budget and FY 1985/1989 Five-Year Defense Program, reportedly projects the navy's budget to rise by 52 percent over the period.[26] This too will no doubt be slashed, but again the other services, the army in particular, will probably have to bear the brunt of Hill cutbacks in the proposed buildup.

The bind on other services' budgets will be increased if, as many believe, the navy's ambitious shipbuilding program proves to be significantly undercosted.[27] Indeed, there is evidence that by using "optimistic estimates" the navy undercosted its original FY 1984 shipbuilding request by some $600 million.[28] When Congress later imposed a 5 percent real growth ceiling on Pentagon spending, the navy reportedly similarly trimmed its FY 1985 ship costs to levels deemed "unrealistically low."[29] Moreover, inflation of shipbuilding costs has been higher than the inflation used for DoD budget purposes.

A readiness and manpower squeeze may result as well. Manning the proposed 600 ship navy will probably take at least 65,000 additional personnel, raising total navy manpower costs by as much as $8 billion.[30] Should the Congress keep holding DoD personnel close to its present ceiling, as it has done for FY 1983 and 1984, this will have to come at the expense of other services' force structure too.

As this sinks in, sharp debate has broken out in the Pentagon over the imbalance of funding between the services. Deputy Secretary of Defense Paul Thayer has reportedly taken the position that the army has been shortchanged and has exchanged sharp words with Lehman over navy spending. Thayer and Undersecretary for Research and Engineering Richard DeLauer also appear to question the cost-effectiveness of several navy programs on technical grounds.[30] Unfortunately, as Dunn and Staudenmeier note, "most of the debate within the Pentagon . . . has concentrated on interservice budget issues rather than on more important strategic questions." They find that it has "all the overtones of a service budget debate with just enough strategic fluff to provide respectability. A comprehensive review of the national security implications is therefore necessary before the options available to policymakers can be appropriately weighed."[32]

The real issue is not whether we need a strong navy — of course we do. We must remain capable of commanding the seas at times and places of our own choosing in order to carry out any force-projection strategy. The issue is rather what kind of navy we can afford for this purpose, given other equally pressing needs. It is a question of priorities for the allocation of constrained resources, and of the likely implications for our strategy of the overall force posture that results.

Current Maritime Strategy

The strategic implications of U.S. commitment to building Lehman's carrier-heavy navy, even at the expense of other capabilities, are painfully clear. As ex-army Chief of Staff Meyer put it, "the issue is do you force yourself in the 1990s to rely on a maritime strategy or do you continue to have balanced forces that can respond in a unified way with all the service elements?"[33] What is really happening is that *the United States is drifting by default toward a primarily maritime strategy.*

If funding of the Reagan administration's ambitious defense buildup is held back by Congress in future years, as seems quite likely, we could end up with a seriously unbalanced, navy-heavy force posture, which in turn will dictate our real strategic options down the road. For one thing, the huge "bow wave" of procurement emphasized in the buildup to date could have serious adverse impact on what is left for readiness and sustainability in future budgets of all the services. For another thing, the big-ship navy will inevitably have to be funded at the expense of adequate ground/air support to our prime overseas allies.[34] As has already happened in FY 1982 to 1983, there will be insufficient funding to complete the prepositioning of army or Air Force equipment in Europe. Creation of flexible rapidly deployable ground/air capabilities for defense of Persian Gulf oil may also be affected. Even other portions of the navy budget, such as antisubmarine warfare (ASW) and amphibious assault shipping, will have to be trimmed if fifteen CVBGs are to remain the core of even a modified naval program, as Secretary Lehman insists will be the case.

That this is the direction in which we are actually going is apparent to others as well. Admiral Turner and Captain Thibault (USN) acknowledge that I am "on solid ground" and justifiably concerned that the U.S. Navy is in fact moving toward a maritime strategy that "would go beyond controlling the seas and would use the seas to project maritime power against the Soviet Union."[35] They further agree that my analysis that the cost of this kind of carrier-heavy navy would in time starve the other services might be "correct," although they go on to suggest a different kind of maritime strategy of their own, as noted later in this chapter.

Thus the maritime school has come into full bloom with the Reagan administration. Although the White House and Defense Department have been cautious about explicating it, the dynamic young Secretary of the Navy, John Lehman, has been admirably forthright. Ignoring Mackinder in favor of Mahan, he insists that geography "overwhelmingly favors the Free World alliance" in today's increasingly maritime world — hence that "maritime strategy is increasingly and self-evidently the *makeweight* of American national security policy."[36] In what he frankly calls the "rebirth of a U.S. naval strategy," he has explained that the navy's carrier-heavy buildup is to give the

United States the ability to "prevail" simultaneously over "the combined military threat of our adversaries" in the Atlantic, Pacific, and Indian oceans via offensive carrier strikes. One reason is that "the growth of the Soviet global navy has eliminated the option of planning for a regionally limited naval war with the Soviet Union. . . . Should conflict begin [between the navies] it will be instantaneously a global naval conflict."[37]

Here we see the first major difference between current naval thinking and that of the Nixon-Ford and Carter administrations — *greater emphasis on roughly simultaneous carrier offensives instead of sequential ones,* thus justifying the fifteen CVBGs already funded — as well as two more the navy requested in its 1985/1989 program.

A second change in emphasis is the stress on offensive carrier operations against land targets, instead of sea control. The navy's demand for fifteen to seventeen big carrier battle groups is not generated primarily by the classic maritime mission of dominating sea lines of communication to protect vital trade, plus overseas reinforcement and resupply. Instead these CVBGs are designed primarily for attack on land targets — among other things to cripple the Soviet navy in its home bases. The maritime school sees this as the best way to maintain sea control and to permit countervailing operations where the Soviets are most vulnerable. The U.S. Navy has long had contingency plans for multicarrier nonnuclear strikes against Soviet naval bases, but clearly these are now being more heavily stressed.

Underlying this maritime school emphasis on early simultaneous carrier strikes is a muted but nonetheless distinct feeling that our allies are weak reeds on which to base U.S. strategy. Their presumed lack of will, as reflected in their reluctance to assume an adequate share of the mutual defense burden, is seen as making it difficult if not impossible for the United States to hold on to Europe or even Persian Gulf oil.[38] Surprisingly, the maritime strategists don't say much about Japan's unwillingness to share mutual defense burdens, by any measure far worse than Europe's. Could this be because Japan lies in a navy-dominated theater of operations?

Sometimes this denigration of allied capabilities is expressed more diplomatically in terms of letting our chief allies assume the responsibility for their own ground/air defense. But it adds up to the same thing. The maritime school tends to hold that the only assured way for the United States to carry any conventional war to the enemy around the fringes of Eurasia is via the sea.

Moreover, it argues vigorously that the United States is too preoccupied with preparing for the least likely case, a Warsaw Pact offensive against the NATO center, at the expense of much more likely contingencies — or even Warsaw Pact nibbles at NATO's vulnerable flanks. Some adherents call for cuts in NATO-oriented U.S. ground and air forces to fund a larger naval buildup.[39] They point out that most wars in the nuclear age have been in the Third World and that even the two major U.S. wars since 1945 have been in Korea and Vietnam (Chapter 7 will analyze the "likelihood fallacy" inherent in this logic). Naturally, posturing primarily to deal with the greater likelihood of Third World conflicts entails greater focus on flexible sea power.[40]

Another feature of current maritime strategy is a fascination with *horizontal escalation*. Rather than attempting to defeat the Soviets where they are strongest, we should adopt a countervailing strategy of attacking them instead in areas where the flexibility of sea power gives us an advantage. If the U.S.S.R. were to attempt to take over Persian Gulf oil, for example, the United States could counter by dealing with such Soviet surrogates as Cuba, South Yemen, Angola, and now — one presumes — Nicaragua, or even sweep the Soviet navy and merchant shipping from the seas. This last would be logical because, as Mr. Lehman says, any direct clash between the United States and U.S.S.R. "will be instantaneously a global naval conflict."[41]

The Sea Control Variant

Although the foregoing is a fair description of the dominant school of maritime strategy, there is a variant that puts primary emphasis on *sea control* as opposed to offensive force projection via carrier battle groups. Its best-known proponents are senior "black shoe" (surface ship) admirals like Elmo Zumwalt (chief of naval operations, 1970/1974), Worth Bagley (his vice-chief); and Stansfield Turner.[42]

In a recent exposition of the case for a more traditional sea-control strategy, Admiral Turner deplores our overemphasis on the European scenario and calls for "greater flexibility for Third World contingencies" via a navy postured more sensibly for sea control. He too attacks spending so much money on the big carriers because they are too expensive, too vulnerable, and too costly to risk in dubious attacks on highly defended Soviet naval bases. Instead, since "sea control is ultimately a war of attrition and losses are inevitable," he favors larger numbers of smaller ships, including smaller carriers. To him, the key value of large carriers is in accommodating big, high-performance aircraft. He argues, however, that technology is driving us away from these to smaller aircraft armed with long-range missiles, which in turn can be carried on smaller ships.[43]

Turner rejoins other navalists in seeing intervention in force "in unexpected and remote areas" as a major mission within a maritime strategy. Because of the uncertainty over where our national interests will be challenged next, "we must build flexibility for the unexpected, which demands cardinal emphasis on capability for forcible entry from the sea, via a greater and considerably revamped Marine Corps amphibious assault capability." He also makes a strong case that smaller carriers are best for this mission as well as for sea control.[44] Although Admiral Turner's navy is preferable to the one the Reagan administration is building, his vague call for "a new military strategy," based on willingness to look at U.S. security needs "with an open mind" smacks of the same old unilateralist naval parochialism which is again in the ascendant.

Admirals Zumwalt and Bagley join Admiral Turner in favoring distributing our sea power over a larger number of smaller ships, rather than spending so much on a smaller number of high-value carriers. However, they place greater emphasis than he does on attacking sea and land targets, not by carrier aircraft but by proliferation of cruise missiles like Harpoon and Tomahawk on existing and new ships, including submarines. They find cruise missiles a much more cost-effective and less vulnerable means of performing this mission, and regard their slow introduction into the fleet as the product of doctrinal conservatism.[45]

What is not clear is whether the sea-control navy advocated by these surface admirals would be less expensive overall than the kind we are now building. Turner claims only that it would not cost any more and would offer compensatory advantages. Zumwalt is undoubtedly correct that proliferating cruise missiles would be much less expensive than employing costly aircraft from big carriers. But he seems to join Turner in believing that the United States needs a much larger navy.

Although Chapter 7 will discuss the flaws in any predominantly maritime strategy, it must be acknowledged here that the maritime advocates do face up to the overriding reality in a context of constrained resources — the necessity for choice. In this sense the maritime-strategy case is more respectable than the JCS consensus that we must have a much larger army and air force as well as navy than are realistically fundable, in order to meet all service needs for a multifront war.

It is also more respectable than navy propaganda that we must meet treaty commitments "to 40 nations," or that we must have some such magic number as "600 ships" to recover the maritime superiority "we have lost."[46] Such numbers games obscure more than they reveal.[47] One has yet to see any convincing analysis, classified or otherwise, that the United States and its allies have "lost" maritime superiority to the Soviet bloc. Although the Soviet navy poses an increasingly serious threat, the U.S. Navy is still far larger in tonnage, qualitatively superior, and far better able (together with our allies) to cope with this threat than budget competition allows its propagandists to admit. Actually, if one counts allied as well as U.S. ships, our total is more than 900 ships, with more than double the tonnage of the ships of the U.S.S.R. and its allies. Naval pessimists also ignore the enormous geographic advantages of our maritime position vis-a-vis that of the U.S.S.R. From a naval viewpoint, the United States has the strategic advantage of "interior lines," and can flexibly shift naval power from one ocean to another far more readily than the USSR.

NOTES

1. See E. R. Lewis, *Seacoast Fortifications of the U.S.: An Introductory History* (Washington, D.C.: Smithsonian Institution Press, 1970).

2. T. Harry Williams, *History of American Wars*, p. 736.   3.   Ibid., pp. 354/355; Russell Weigley, *American Way of War*, pp. 228/232.

4. Weigley, *American Way of War*, pp. 228/229.

5. Dean Acheson, *Present at the Creation* (New York: Macmillan, 1969), pp. 491/492.

6. Ibid., p. 373.

7. However, navy opposition to JCS reform and a more integrated military staff system is not the only important reason why it has so often been sidetracked. As R. K. Betts notes in "Conventional Strategy: New Critics, Old Choices," *International Security*, Spring 1983, p. 155, Congress "normally prefers to constrain the executive [branch] by dividing it and to deal directly with the separate services. The putative fear of a Man on Horseback is more a rationale for maintaining legislative clout than for protecting the nation."

8. H. Stimson and McG. Bundy, *On Active Service* (New York: Octagon Books), p. 506.

9. W. W. Kaufmann *Planning Conventional Forces*, pp. 11/13, describes how the navy was never willing to allocate naval forces to specific scenarios in the various force-sizing studies made by different administrations, but preferred to rely on "flexibility."

10. In the nuclear field, DoD's refinement of nuclear strategy toward a wider range of options culminated in Presidential Decision #59 of 1980.

11. For example, James D. Hessman in "Sea Power and the Central Front," *Air Force Magazine*, July 1983, p. 52, notes how "the U.S. Navy was not, in the late 1970s, ardently supportive of what came to be called 'the Central Front Strategy' developed under then Secretary of Defense Harold Brown. As the navy saw it, this strategy would result in reduced funding

for navy and marine corps units around the world in order to build up and modernize U.S. air and ground units assigned to the Central Front in Europe."

12.   James Abellera and Rolf Clark, "Forces of Habit," pp. 2-3. The numbers cited are in constant 1981 dollars and exclude marine corps funding. Contrast this to the repeated claims of Navy Secretary Lehman about how the previous administration cut the navy, or the allegations in Jeffrey Record and Robert J. Hanks, *U.S. Strategy at the Crossroads* (Washington, D.C.: Institute for Foreign Policy Analysis, July 1982), p. 4.

13.   See Harold Brown, *Thinking about National Security* (Boulder, Colo.: Westview Press, 1983), pp. 174/178, for a trenchant discussion of this issue.

14.   The best source of administration strategic thinking lies in the annual Pentagon posture statements. The *DoD Annual Report to the Congress, Fiscal Year 1983*, was Weinberger's first. At the beginning (p. I-3), it contends that "our defense policy has not only become obsolete because of new threats to our security, it has also been discredited by its failure to recognize and cope with the deterioration in the global military situation. In fact, obsolete strategic concepts have stood in the way of necessary reforms." Note the contrast with former Secretary Brown's emphasis in *Thinking about National Security* on the essential continuity of national security policy over many administrations.

15.   Ibid., p. III-91.

16.   Ibid., pp. I-16, I-17. I have never found a real "short war theorist." Instead the issue has always been one of priorities. Given constrained resources we must posture first for deterrence and initial defense, on the grounds that if deterrence fails and the initial defense crumbles, the issue of longer war in the areas lost becomes academic. In fact, fiscal constraints have compelled the Reagan administration to give little higher priority to sustained defense than have previous administrations.

17.   See for example, Barry Carter in *Rethinking Defense and Conventional Forces*, pp. 20, 29, and Record and Hanks, *U.S. Strategy at the Crossroads*, pp. 25/27.

18.   George Wilson, "U.S. Defense Paper Cites Gap Between Rhetoric, Intentions," *Washington Post*, May 27, 1982, p. A-4. 19. Citations are from a speech by Judge William Clark, Assistant to the President for National Security Affairs, at the Center for Strategic and International Studies, Georgetown University, Washington, D.C., May 21, 1982. It is the most authoritative explication of the Reagan administration's strategy review.

20.   *FY 1983 DoD Posture Statement*, p. I-30, repeated in *FY 1984 DoD Posture Statement*, p. 46.

21.   Comment by Former Secretary of Defense Melvin Laird.

22.   Lehman himself estimated the cost of a new CVBG at $16.8 billion in FY 1983 dollars, which means that three more CVBGs would total over $50 billion in capital costs alone.

23. "Too Late To Stop Navy Buildup, Says Navy Secretary," *Washington Post*, December 2, 1983, p. A-6.

24. Although navy investment was swollen in FY 1983 by funding for two new carriers in one year, in fact over the last decade (FY 1974/1983) the navy's investment budget has averaged considerably *more* than double that of the army.

25. See Deborah Kyle in *Armed Forces Journal*, August 1983, pp. 8/12, for a fascinating series of articles on how "Army Procurement Always Bears Brunt of SASC Cuts," "Army Gets Smallest Increases When House Adds To Service Hardware Funds," "House Appropriations Committee Cuts Navy $ Most While Senate Panel Trims Army $."

26. Richard Halloran, "Military Sees Big Rise in Funds for '85," *New York Times*, October 20, 1983, p. B-11.

27. See Franklin Spinney, "Defense Facts of Life," December 1980, an in-house study released by DoD, p. 37; or David Wood, "Defense Costs Held to Exceed Huge Budget," *Los Angeles Times*, March 5, 1983, p. 10.

28. "Navy Lowered Its Experts' Cost Estimates To Make Ship Budget More Optimistic," *Wall Street Journal*, March 28, 1983, p. 4. It quotes a memo by Assistant Navy Secretary George Sawyer directing that this be done.

29. Charles Corddry, "Pentagon Split by Battle over Navy's Budget," *Baltimore Sun*, August 17, 1983, p. 1, citing Deputy Secretary of Defense Thayer's views; and George Wilson, "Lehman Wins a BudgetBattle," *Washington Post*, September 8, 1983, p. 1.

30. *Manpower for a 600-Ship Navy,*, (Washington, D.C.: Congressional Budget Office, August 1983). Commander M. B. Edwards (USN), in "Supporting The Six Hundred," (Naval Institute Proceedings, August 1983) finds (p. 51) that considerably more than the number of planned underway replenishment groups will be needed to support fifteen CVBGs.

31. Ibid. See also Fred Hiatt, "Feud Erupts on Navy's Future," *Washington Post*, October 11, 1983, p. 1; Charles Corddry, "Pentagon Officials Warned to Curb Feud," *Baltimore Sun*, October 12, 1983, p. 8.

32. K. Dunn and W. O. Staudenmeier, "Strategy for Survival," *Foreign Policy*, Fall 1983, p. 24.

33. "Pentagon Regroups for the Leaner Years," *Washington Post*, August 14, 1983, p. A-12.

34. Paul Nitze and Leonard Sullivan described the process in *Securing The Seas,* an Atlantic Council Policy Study (Boulder, Colo.: Westview Press, 1979), pp. 323/326. The majority of the study group opposed such shifts in resources.

35. Stansfield Turner and George Thibault, "Preparing for the Unexpected: The Need for a New Military Strategy," *Foreign Affairs*, pp. 123 and 134. See also the rebuttals and sur-rebuttals by them, Secretary Lehman and myself in *Foreign Affairs*, Winter 1982/83, pp. 453/457.

36. Remarks to the Navy War College Current Strategy Forum, New-

port, Rhode Island, June 21, 1983, pp. 9 and 12; see also "Rebirth of a U.S. Naval Strategy," *Strategic Review*, Summer 1981, esp. pp. 13/14.

37.   Posture Statement of Secretary Lehman before the House Armed Services Committee (HASC), February 8, 1982, p. 6; see also "Rebirth of a U.S. Naval Strategy," *Strategic Review*, Summer 1981, pp. 9/15.

38.   Although this view is seldom voiced officially, it is expressed informally. For the views of outside critics, see, for example, Record and Hanks, *U.S. Strategy*, pp. 1/15. 39. In ibid., p. 15, Record candidly calls for sharp cuts in the U.S. Army to free up funds for sea-based forces. So too do William Lind, Norman Polmar, and Dominic A. Paolucci, three dissenting members of the Atlantic Council Working Group which produced *Securing the Seas*, pp. 445/446.

40.   Turner and Thibault make this case more elegantly than the navy itself in "Preparing for the Unexpected," pp. 122/135. 41. HASC Hearings on FY 1983 Defense Authorization, No. 97-33, pp. 561/562.

42.   Dunn and Staudenmeier, "Strategy for Survival," pp. 24/26, discern three different U.S. variants: the "Lehman school"; the "manipulation school"; and the "unilateralists," epitomized by Jeffrey Record. Even Record, however, wants a sea-control navy (*U S Strategy*, pp. 33/34).

43.   Turner and Thibault, "Preparing for the Unexpected," pp. 125/129.

44.   Ibid., pp. 129/132.

45.   See, for example, Elmo Zumwalt and Worth Bagley, "Military Doctrine Old and New; Conservatives Retain the Upper Hand," *Washington Times*, February 21, 1983, p. 2-C.

46.   Interview with Secretary of the Navy John Lehman, *Military Science and Technology*, June 1982, p. 26.

47.   Abellera and Clark, "Forces of Habit," pp. 44, 48/51, point out that in the past the U.S.S.R. deliberately followed a small-ship policy which has resulted in impressive numbers, whereas the United States has followed an opposite policy. However, they see the Soviets too now shifting toward larger ships of greater capability and sustainability, which could mean a decline in total ship numbers just as occurred in the United States during the postWorld War II years.

CHAPTER 7

Flaws in the Maritime Strategy

It is the central thesis of this book that, even though the maritime school faces up to the necessity for choice, its choice is a wrong one. Cutting the coat to fit the cloth by concentrating on a primarily maritime strategy cannot adequately protect our vital interests in Eurasia because it cannot adequately deter a great land-based power like the U.S.S.R. It is an unbalanced strategy, one that ignores geopolitical realities, as this chapter will seek to demonstrate.[1]

The Fatal Flaw

First and foremost, could command of the sea and carrier strikes around the periphery of the Soviet homeland deter or prevent the U.S.S.R. from overrunning Europe and the Middle East oil fields, cowing or emasculating China, or mounting a land-based missile and air threat to nearby Japan that would dwarf Hitler's 1944 V-1 and V-2 threat to wartime Britain? The answer is no. Offensive naval operations simply could not prevent the projection of Soviet power outward in Eurasia by air and land.[2] Even if all Soviet home and overseas naval bases were put out of action, and Soviet naval and merchant vessels swept from the high seas, this would not suffice to prevent Moscow from seizing or dominating the rimlands of Eurasia, including the two great industrial agglomerations of Europe and Japan, and cutting off their economic lifeblood — Middle East oil. Thus a peripheral maritime strategy gives up the chief prize of any U.S.-Soviet global contest — the resources of Eurasia.

Nor could a countervailing strategy of conventional naval attacks on land targets impose remotely comparable damage on a great Eurasian land power like the U.S.S.R. Even if carrier strikes around the Soviet periphery could be carried out with great success with acceptable losses, it would be like sticking pins in the hide of an elephant. Overpowering Soviet overseas surrogates like Cuba, South Yemen, Ethiopia, Vietnam, or Libya also could not produce even remotely countervailing value (any more than seizing France's overseas possessions brought Napoleon to heel). As for cutting off overseas trade, the autarchic Soviet economy is by no means as dependent on it as are those of the United States, Europe, and Japan.

Thus a predominantly maritime strategy would offer little hope of being able to prevent a decisive shift in the balance of power against the United States and any remaining allies. It is a pity that maritime strategists — whether in Whitehall or in the Pentagon — so often ignore such basic balance-of-power considerations.

Could Europe and Japan Go It Alone?

Some advocates of a maritime strategy evade this issue by contending that Europe is now more than rich enough to provide for its own defense — or at least for its own ground forces. Presumably the same is also true of Japan. These advocates go further and suggest that major U.S. ground force withdrawals would stimulate our allies to do more for their own defense, thus leading to more rational burden-sharing as the United States shifted greater resources to naval purposes.[3]

True, a stronger and more integrated European defense effort would be highly desirable, and has always been favored by Washington on strategic grounds. But to suggest that Europe alone would do what Europe plus America so far have not is whistling in the wind. The conventional NATO/Warsaw Pact balance is already so unsatisfactory that substantial U.S. troop withdrawals would involve highly imprudent risks. If deterrence is the name of the game in a nuclear age, then substantial U.S. forward deployments surely enhance it more than posturing to return to Europe when it may be too late. Nor is it likely that our European allies would see fit to spend a great deal more under these circumstances, when the United States would be contributing so much less.

Far more likely would be a desperate move (led by France) toward development of a European nuclear deterrent, together with powerful pressures for accommodation with the USSR. This is no doubt equally true of Japan and China in East Asia should the United States seem to be pulling back from them. In short, building up a more credible conventional deterrent would not be facilitated by decreasing the U.S. contribution; the consequences would more likely be the opposite.

Undermining Our Alliances

Indeed, adoption of a predominantly maritime strategy would have a devastating impact on the very network of alliances on which the United States is so dependent to maintain a credible deterrent or defensive balance vis-a-vis the U.S.S.R. Some eroding of alliance cohesion is already apparent as the Reagan administration has fallen into sharp differences with our allies. It has also deemphasized most of the NATO initiatives launched by its predecessor, causing a serious loss of momentum in the effort to achieve more rational burden-sharing.

Moves toward primary reliance on a maritime strategy and posture would further the process of erosion. Our chief allies would quickly perceive its implications, particularly if budget constraints compelled us to write off as unsustainable our land/air commitments to the defense of Europe and Persian Gulf oil. Hence few allies would welcome a U.S. maritime strategy aimed primarily at naval dominance — even if it protected their own overseas trade — if the price were to expose them to defeat at home. Our restive allies, already concerned over the declining credibility of the U.S. nuclear umbrella, would correctly perceive such a U.S. strategy as either a form of U.S. global unilateralism or a form of neoisolationism — a cutting back on U.S. commitments abroad.

Moreover, a strong case can be made that the United States could not even conduct a successful "countervailing strategy" of sweeping the Soviets from the oceans and bottling them up in the narrow seas without active cooperation from many of our allies. Barring the Dardanelles would require active Turkish and probably Greek cooperation. Clearing the Mediterranean and keeping it cleared would require active French, Spanish, Italian, and Greek help. Closing the Baltic exits would be difficult without Norwegian and Danish assistance. Keeping the Red Banner Pacific Fleet penned up in the Sea of Okhotsk would require the active cooperation of Japan. Loss of Norway would make it doubly difficult to launch strikes at the Kola Peninsula. Yet their cooperation would be doubtful if the United States no longer contributed to their ground/air defense.

The Likelihood Fallacy

Nor can the United States afford to posture primarily for the most likely contingencies at the expense of the most critical ones. True, the main contingencies against which the United States has postured are the least likely. True, the greatest likelihood of conflict is in the volatile Third World. This has been the case since the dawn of the nuclear age. Because our resources are limited, however, we cannot afford to fall prey to this *likelihood fallacy*.

It is important to remember that the most adverse strategic consequence of the Vietnam War was its enormous diversion of resources from our global defense posture and the expiatory cutback in defense spending in its aftermath. The "loss" of the Indochina peninsula itself was hardly a body blow to the U.S. strategic position worldwide. Similarly, the "loss" of Cuba, Angola, Ethiopia, or Nicaragua cannot be said to have undermined our strategic position, however much these losses may have inconvenienced us. This is not to suggest that the United States should ignore third-area conflict, only that our commitments should not be allowed to outrun our interests, as happened in Vietnam.[4] Indeed, a legacy of Vietnam has been to make it more difficult for the U.S. Congress and public to contemplate limited interventions than a major U.S.-Soviet conflict — witness the brouhaha over Lebanon and El Salvador. Our strategy must take into account this political constraint.

Moreover, a major reason why U.S.-Soviet conflict in Western Europe or Northeast Asia remains unlikely is precisely that we and our allies have devoted great efforts to deterrence in these vital areas. Deterrence is the name of the game in a nuclear age. Just because the likelihood of direct attack on our most vital interests is relatively low is no reason for not continuing to invest heavily in keeping it low. America has survived the "loss" of Cuba, Ethiopia, Vietnam, and so on. Could it survive as well the loss of Western Europe, Japan, or Persian Gulf oil? The *reductio ad absurdum* of the likelihood fallacy is that, since nuclear conflict is the least likely of all contingencies, we need not spend so much on deterring it. In short, we must relate our conventional forces primarily to a strategy designed to preserve our vital interests, if necessary at the expense of much else. This cannot be achieved by means of a peripheral maritime strategy (see Chapter 8 for discussion of specific regional cases).

Would Horizontal Escalation Work?

A parallel case can be made against *horizontal escalation*. The trouble with such forms of countervailing strategy as seeking to hold Cuba, South Yemen, Ethiopia, Angola, or Vietnam hostage if the U.S.S.R. threatened to take over Persian Gulf oil, is that the relative values are wholly incommensurate. The same has already been suggested with respect to driving the Soviets from the seas. Indeed, it is hard to find any feasible countervailing gains that would compensate strategically for the loss of Persian Gulf oil, Western Europe, or Japan. There are only consolation prizes.

On the contrary, we must examine carefully whether horizontal escalation would not be more to Soviet strategic advantage. True, the United States would have distinctly superior ability to extend any conflict to places like Cuba, Nicaragua, Ethiopia, South Yemen, or Angola, given its superior force-projection capabilities and ability to interdict any Soviet support. Horizontal escalation, however, is a game that two can play. Indeed, the U.S.S.R. has superior capabilities to escalate by land and air against not only NATO Europe, but also the Persian Gulf region, Korea, Pakistan, or even China. Unfortunately, most areas lying under Soviet threat are much more important to our side than the U.S.S.R.'s overseas surrogates are to theirs.

Thus horizontal escalation could prove to be an enormous diversion of scarce military resources at the expense of far more essential wartime missions or U.S. ability to deter. Moreover, there are less costly means of coping with Soviet surrogates (see Chapter 8).

Here, however, is an issue that repays differentiation between declaratory strategy and actual wartime moves. U.S. threats of horizontal escalation certainly complicate Soviet planning. Such threats also help deter Soviet surrogates from aggressive action since the United States is clearly able to put Soviet surrogates at risk. Certainly we want potential foes to worry about this possibility. But this is far different from imprudently erecting such a peripheral strategy into something that could actually achieve adequate countervailing value to moves open to the U.S.S.R. This is a mirage.

Automatic Worldwide War?

Serious questions also need to be raised about the shibboleth that a direct U.S.-Soviet military clash in any region would almost certainly lead to a global conflict, at least a conventional one. Admittedly, this is a risk against which we must hedge. It also tends to justify a strategy of horizontal escalation and a multifront force posture. However, it is important to ask whether it would be to the strategic advantage of either side. Above all, any such widespread conflict would immediately increase the likelihood of nuclear escalation. Crossing the nuclear threshold would create for Washington as well as Moscow threats of homeland devastation besides which other strategic calculations pale. This, of course, is why nuclear deterrence has apparently been so effective where perceived vital interests were at stake. In practice, the two nuclear superpowers have carefully avoided direct military confrontation (see Chapter 1).

The U.S.S.R. has been extremely careful to use surrogate forces in Korea, Vietnam, Angola, and Ethiopia, and to attempt to keep such conflicts strictly limited.[5] The United States, even though it intervened directly in Korea and Vietnam, was in turn careful to limit the scope of these conflicts. We learned a lesson in Korea when General MacArthur ignored Beijing's warnings that it would intervene if U.S. forces went to the Yalu River, thus threatening Manchuria. In Vietnam we leaned over backward, some say too far, in our effort to minimize the risk of Chinese intervention. We accepted limited or in some cases total sanctuaries for the enemy. Nor should we forget that Soviet surrogates in turn were equally careful not to violate U.S. sanctuaries like the Philippines or Thailand. Indeed, what is remarkable about most local conflicts in the nuclear age is the care exercised by the participants and their backers to limit the risks of escalation.

Nor is it at all clear that the U.S.S.R. would see strategic advantage in promptly expanding any direct U.S.-Soviet clash into a global war. Aside from greatly increasing the risks of nuclear escalation, it would disperse Soviet forces instead of concentrating them. If the Warsaw Pact forces attacked NATO, for example, why should the U.S.S.R. simultaneously initiate hostilities in other regions?[6] Would this divert enough U.S. forces to ensure that Pact forces could conquer continental Europe? Most analyses suggest that the Pact could defeat NATO in Europe right now, whatever the United States did short of nuclear escalation.

Therefore, why shouldn't the U.S.S.R. hold off attacks elsewhere to minimize risks and complications until it had won decisively in one key theater? At that point Moscow could shift overwhelming force to cow or defeat potential foes in other theaters with high likelihood of success. Let us not forget that it was Moscow that preserved studied neutrality vis-a-vis Tokyo in 1940/1945 until Germany had been defeated. If Moscow moved instead into the Persian Gulf, why simultaneously take on NATO instead of merely mounting a threat-inbeing to pin down NATO forces? Surely NATO would not retaliate by attacking the USSR.

It is equally hard to see how the United States would find it strategically advantageous to initiate global hostilities in event of a Soviet regional attack. Conventional carrier strikes against the Soviet homeland surely could not hurt the U.S.S.R. enough to offset the risks. As suggested in the case of horizontal escalation, expansion of hostilities could well be of net disadvantage to the United States, even if it could get its chief allies to join in — a dubious proposition (see Chapter 8). One price we pay for a coalition policy and posture is that we are less able to function independently.

Another line of argument is that expansion of U.S.-Soviet hostilities anywhere would rapidly escalate to global war because the naval forces and merchant shipping of the two contenders are inextricably mixed all over the world. What if a surface warship perceived itself as coming under submarine attack, or vice versa? Don't we have to allow our ships the inherent right of self-defense? In other words, we could not prevent World War III because global maritime escalation is inevitable.

Given the awesome risks involved, this line of argument is simply unacceptable. If positive control over every ship at sea is necessary, so be it. In a nuclear age no nation can afford to let global war be triggered by the action of individual ship captains when much larger issues are at stake. Rules of engagement designed to prevent inadvertent expansion of any conflict are essential in the conventional as well as nuclear arena.

Denigration of Sea Control

Owing to budget constraints, our navy's current emphasis on offensive force projection is also having the effect of detracting from investment in sea control. The new CVBGs are so expensive, with their panoply of escorting Aegis ships and destroyers plus their expensive air groups, that other navy programs have to give. The FFG-7 class of ASW frigates has been closed out in the FY 1984 budget. Instead, the navy is going to build a highly expensive class of DDG-51 Aegis destroyers. Moreover, attacks on heavily defended land targets call for big carriers, whereas sea control could be exercised with smaller carriers.[7]

In fact, several recent analyses suggest that protection of the Atlantic convoy routes against the growing Soviet bomber threat could best be carried out with a combination of AWACS and long range fighters with long-range missiles, land-based on islands or on continental bases fronting the eastern Atlantic. Yet to date little has been done to carry out this concept, perhaps because the navy fears it might lose the mission to the air force. If this is indeed a major reason, then the solution is straightforward: Give the navy responsibility for it and let it have its own AWACS and F-15s.

Vulnerability of Carriers

Last but not least are serious questions about the vulnerability of carriers, especially when they are used against highly defended land targets like Soviet naval bases. Turner and Thibault assess at length the difficulties of achieving tactical surprise and likely repeated attacks on carriers by land-based bombers, missiles, and submarines.[8]

Harold Brown points out how carriers can be located by satellite reconnaissance and how in time they will probably become vulnerable to land-based missile attack guided by satellite.[9] If high-value carriers are exposed to such risks, admirals will be reluctant to use them for risky attacks. Despite all the bold talk about attacking the Kola Peninsula or Vladivostok, Admiral Turner avers that he "has yet to find one admiral who would even attempt it."[10]

A further problem is that CVBGs have to expend too much on self- defense to have much offensive punch at longer ranges — a ten- to twelve-plane squadron or two per carrier, currently A-6s carrying "dumb bombs." There has been "an inexorable trend in the last few years to fewer attack aircraft on our carriers as the need for defensive and support aircraft — such as early warning, antisubmarine and tankers — has increased."[11] Out of the estimated $17 billion total cost of a new CVBG (in FY 1983 dollars), less than 10 percent appears to go for longer range offensive systems like A-6s (A-7s lack the range for attacks on heavily defended land targets).

Are carrier-launched bombers in any case the optimum mode of attack for heavily defended land targets? Former senior admirals like Elmo Zumwalt argue that proliferating long range cruise missiles like nonnuclear Tomahawk provides a better, cheaper, and less vulnerable mode of attack. That the current navy leadership is also persuaded of their efficacy is suggested by its programs to put over a thousand Tomahawks on submarines, cruisers, and even four refurbished battleships. If a renovated battleship can carry some 350 of these missiles, as is planned, it seems much more cost-effective than a CVBG. Moreover, instead of attacking Soviet naval bases, we could effectively bottle up Soviet fleets with submarines, and with choke point defenses such as mines.

In sum, the kind of carrier-heavy navy we are building, and the peripheral maritime strategy for which it is designed, cannot meet our basic strategic needs. Even if we simultaneously swept the Soviets from all the seven seas at the outbreak of a war, this could not alone prevent the U.S.S.R. from dominating the entire Eurasian landmass, including such vital areas as Europe, Japan/Korea, and the Persian Gulf oil fields. Only land and air power as well could do that. Nor could strategic diversions in pursuit of horizontal escalation alter this basic reality. Moreover, the current costly emphasis on big carriers for offensive force projection is even eroding our ability to perform the essential naval mission of sea control.

True, carriers still serve many useful purposes. They are splendid for Third World conflict. But we already have thirteen of them (with another building), plus five big marine helicopter/VSTOL carriers (about 40,000 tons), which have a secondary mission of sea control.

Thus we can maintain a large force of at least twelve big carriers through the 1990s even if we never build another. They are sufficient for a balanced strategy provided that we can use the inherent flexibility of sea power to conduct sequential operations, instead of near simultaneous carrier offensives in several theaters.[12] Thus, at a time when Congress is cutting back the Reagan defense program, we cannot afford to put so many eggs in the big carrier basket at the expense of a balanced posture to execute a balanced strategy aimed primarily at preserving our vital interests abroad. To this we now turn.

NOTES

1.   For an analysis of geopolitics in the contemporary international context, see Colin S. Gray, *The Geopolitics of The Nuclear Era: Heartlands, Rimlands, and the Technological Revolution*, (New York: Crane Russak, 1977).

2.   As Betts puts it, "all the aircraft carriers in the world could not stop the Soviet army from overrunning Europe." R. K. Betts, "Conventional Strategy: New Critics, Old Choices," in *International Security*, Spring 1983, p. 140.

3.   See for example Jeffrey Record and Robert J. Hanks, *U.S. Strategy at the Crossroads* (Washington, D.C.: Institute for Foreign Policy Analysis, July 1982), pp. 30/32, 57. Record would leave 60,000 men in U.S. tactical air forces in Europe to demonstrate our continued commitment to NATO. See also Stansfield Turner and George Thibault, "Preparing for the Unexpected: The Need for a New Military Strategy," *Foreign Affairs*, Fall 1982, p. 133.

4.   The Western Hemisphere is a special case. Admittedly, keeping the Americas from being exploited by the U.S.S.R. or its surrogates is perceived as a key element of our security policy; but this is not a strategic mission that need take a large proportion of U.S. forces.

5.   The 1962 Cuba Missile Crisis was a direct confrontation, but fortunately no shots were exchanged.

6.   Similarly, one might ask why the U.S.S.R. should attack on both of NATO's flanks simultaneously with attacking the center. A decisive Soviet march to the Channel in the Center Region would leave both flanks totally vulnerable to takeover, as was the case when Germany won the battle for France in 1940.

7.   Turner and Thibault, in "Preparing for the Unexpected," p. 125, point out that sea control calls for navies to "distribute their power and value over as many ships as possible rather than concentrating them in just a few." As the axiom goes, one ship cannot be in two places at the same time.

8.   Ibid., pp. 125/129.

9.   Harold Brown, *Thinking about National Security* (Boulder, Colo.: Westview Press, 1983), pp. 176/177.

10. Letter in *Foreign Affairs*, Winter 1982/83, p. 457.

11. Turner and Thibault, "Preparing for the Unexpected," p. 127.

12. W. W. Kaufmann also takes the view that twelve modern carriers are enough in "The Defense Budget," in *Setting National Priorities* (Washington, D.C.: Brookings Institution, 1982), p. 91.

CHAPTER 8

A Sound Coalition Strategy

If maritime supremacy amounts to a no-win conventional strategy, because it tends to sacrifice the very allies on whom our own security is so heavily dependent, does rejuvenating our alliances offer enough greater promise to be a preferable alternative? Some critics tend to discount this as a traditionalist or orthodox Atlanticist approach.[1] But is a strategy outdated simply because it is of long-standing, or because it still looks to defending Western Europe as a vital U.S. interest?

Obviously, it would be strategically desirable from a U.S. view-point to have a strong collective defense of such vital Eurasian areas as Europe, Northeast Asia, and the Persian Gulf oil region. Failure to hold any one of these three key areas could ultimately result in a decisive shift in the global balance of power against the United States. Helping our allies to defend these areas clearly requires balanced ground and tactical air forces as well as command of the sea to permit us to deploy and support them. Thus the "coalition" or "continental" school tends to see our present strategy as sound. We have no other strategically sensible choice. But this is not good enough. The crucial flaw in the coalition approach, as acknowledged in Chapter 3, is that *the United States and its allies have never developed a coalition posture adequate to execute it at politically acceptable cost.* However desirable the strategy, it still falls short on the resource side of the strategy-resource mismatch.

The Need for a New Consensus.

For one thing, *there is as yet no Alliance consensus on the proposition that the declining credibility of nuclear deterrence dictates greater reliance on conventional deterrence/defense.* To the contrary, our NATO allies, who have always tended to favor primary reliance on nuclear deterrence (see Chapter 1), still cling to a version of flexible response that allows for only a brief conventional pause. Indeed, allied governments see the U.S. shift toward greater emphasis on conventional capabilities as dangerously eroding nuclear deterrence. For this reason they have reacted sharply to such notions as "no first use."[2] In part this reflects the longstanding European preference for defense on the cheap via a primarily American nuclear umbrella. Similar preferences exist in Japan.

A new incentive is coming into play, however. The strong popular opposition to NATO's deployment of intermediate-range nuclear cruise and ballistic missiles suggests that Europeans too are increasingly beginning to question what James Schlesinger calls "the fatal flaw in the Western Alliance system . . . its over-reliance on nuclear deterrence."[3] As the allies come to realize the need for a stronger conventional deterrent at a time of nuclear stalemate, it should be possible to fashion a better consensus to this end. At the very least the United States should keep pressing this argument with its allies.

If a coalition approach remains indispensable to both the United States and our allies, there is a strong political case for using it as a new common rallying point for the alliance.

> Transatlantic disaffections, sturdy perennials since the turn of the decade, continued to sprout luxuriantly throughout 1982. They have been nourished by two as yet inchoate forces which, if unchecked, will logically lead to the end of the alliance: the trends toward neutralism in Europe and toward unilateralism in America.[4]

This book is not the place to discuss the well-known political and economic frictions that many see as undermining NATO. Suffice it to say here that failure to adapt its strategy and posture to the realities of nuclear stalemate, though perhaps avoiding added frictions in the short run, could even more seriously erode NATO's deterrent credibility — hence its very reason for being — over the longer run. Past experience also suggests how difficult it will be to get our allies to spend enough to achieve credible conventional deterrence. Getting them to spend it more efficiently may be even more difficult (see Chapter 9). Given the real obstacles entailed, building up sufficient strength to offer high-confidence conventional deterrence/defense of all key areas of Eurasia will be at best a long- term proposition.

Meanwhile, is it possible to modify our present operational strategy itself to enable it to be executed with higher confidence within the resources likely to be available? To the extent that we can make our strategy more realistic, it should help justify the necessary outlays. Therefore, exploring at how we might adjust the strategy side of the strategy versus resources mismatch is the main subject of this chapter.

Forestalling Multifront War

One option is to reevaluate the "worst-case" assumption that any direct U.S.-Soviet regional clash would almost automatically lead to global conflict.[5] In Chapter 7 the case was made that *both* superpowers would prefer to localize any local conflict if feasible. First and foremost, any direct U.S.-Soviet nonnuclear clash would greatly increase the risk that one side or another would feel compelled to cross the nuclear threshold. If so, globalization of such a clash would by definition increase such risk much further. Against this risk, what commensurate strategic interest would either side seek to serve?

Take for example a U.S.-Soviet clash over Persian Gulf oil. Chapter 7 has already dealt with some of the flaws in a countervailing strategy of horizontal escalation. Would getting our major allies in Europe or the Far East to join us in expanding a Persian Gulf conflict to those theaters be of net strategic advantage to us? If we could cause the USSR to suffer commensurate conventional losses in those theaters, there might be at least a strategic case for doing so. In neither case, however, does the Western coalition have sufficient offensive capability to put commensurate Soviet interests at serious risk. Indeed, our present *defensive* capabilities are inadequate. Thus widening the war would be strategically disadvantageous to our side, making it highly doubtful that our NATO or East Asian allies would be willing to join us in expanding the conflict despite their dependence on Persian Gulf oil. Without their active participation, how could we successfully coerce the U.S.S.R.? With naval pinpricks of no decisive value around the Soviet periphery?

On the contrary, it is the U.S.S.R. that might achieve strategic advantage by widening a U.S.-Soviet regional war in the Persian Gulf region. At present it has a greater capability to do so in Eurasia. To this, two important caveats must be attached. One, already mentioned, is that this would greatly increase the perceived risks of nuclear escalation. The other is that the U.S.S.R. need not attack everywhere initially to achieve decisive results. If Moscow could successfully dominate the Persian Gulf, then the oil supplies on which

Europe, Japan, and the Third World depend so heavily would be available only at Soviet sufferance — a powerful instrument of political coercion. Moreover, the U.S.S.R. would be in no worse position to extend hostilities later to other theaters. It could employ a strategic doctrine of sequentialism rather than simultaneity — as it did in seeking to defeat Germany in World War II before taking on Japan.

In the last analysis, the most likely circumstances in which global expansion of a regional conventional conflict might occur would be if one side perceived that the other intended to expand it, hence chose to preempt. Surely there are many ways in which one side could signal the other credibly that this was not the case. Surely there are also many means of going to a higher state of defensive readiness. Such measures as precautionary deployments, reinforcement of vulnerable areas, even partial mobilization, would reduce the likelihood of a wider war, while still not being unduly provocative. Without spelling out all the possibilities, we clearly need to examine how best to forestall a multifront war.

Can Europe Be Defended Conventionally?

Let us now turn to the question of whether there are useful strategic alternatives to our present coalition strategy and posture in the key regional theaters. To start with, before writing off defense of Europe in favor of a peripheral maritime strategy, it seems worthwhile to assess whether high-confidence conventional deterrence/defense can be achieved at reasonable cost. Here it is important first to define what we mean. NATO's flexible-response strategy hardly calls for it to sustain a protracted conventional war lasting many months if not years. Aside from the enormous peacetime costs of such preparedness, it would certainly tend to undermine the credibility of nuclear deterrence, on which NATO still principally relies. Moreover, with their memories of the destructiveness of World War II still vivid, our European allies are fearful of a replay that over time could cause casualties and damage comparable to a nuclear exchange.[6] Realistically, then, NATO must continue to stress deterrence — as the best way to forestall in the first place a conventional as well as a nuclear war that no one could really win.

Thus NATO's greatest need is not to change its strategy, although modifying MC 14/3 to clarify how long we should seek to defend conventionally would be desirable. Rather, it is to generate a sufficiently strong initial conventional defense — capable of lasting weeks instead of days in order to outlast a Soviet blitzkrieg — to maximize deterrence. As Francois de Rose puts it, "the key task is to create the capabilities that can lift from the Alliance . . . the incubus of early resort to nuclear weapons in order to avert certain defeat on the conventional battlefield."[7]

Warsaw Pact forces are not postured for a protracted conventional war either, but rather for a quick blitzkrieg win. To deny them this option without decisive loss of territory would enormously complicate Soviet planning. Historically speaking, general staffs do not plan long wars of attrition; the attacker invariably seeks decision in the opening campaign, aware that if he fails all else becomes uncertain. To do otherwise, especially in a nuclear age, is just too risky. Moreover, a stalwart conventional defense makes crossing the nuclear threshold if it fails that much more credible.

Yet is even *initial* conventional defense of NATO Europe feasible at reasonably affordable cost? Against a formidable Warsaw Pact threat, NATO at present can only put up a thin forward defense without much depth to absorb an armor-heavy breakthrough. Doing better will inevitably cost more, and the next chapter will assess the problems of resource generation. This chapter will discuss a number of military options for strengthening initial antiblitzkrieg defense, some of them quite low in cost.[8]

One such option would be to abandon the effort to hold well forward in favor of trading space for time, conducting a fighting retreat back to a defensible position like the Rhine-Yssel River line in the expectation that sufficient reinforcements could arrive by then to hold it. This was the preferred U.S. strategy back when NATO deployed too few peacetime covering forces to defend well forward and when U.S. reinforcements could arrive only belatedly by sea. Such a strategy, however, was abandoned as politically as well as militarily unsound.

Politically, the Germans could hardly accept a strategy that condemned them to initial loss of the bulk of German territory and industry, which lie beyond the Rhine. Therefore, Bonn insisted, as the price for its crucial contribution to NATO, that a forward-defense strategy be adopted. This remains a German political imperative for wholly understandable reasons.

The prospect of twelve heavy Bundeswehr divisions, plus U.S. and French reinforcements, also made forward defense appear to be militarily feasible in time. However, France's later withdrawal from NATO's military structure and its emphasis on an independent national nuclear deterrent at the expense of its conventional contribution were a blow to forward defense.

Nor does forward defense actually mean positional defense with no yielding of ground, as opposed to tactics of maneuver. Obviously, preselected and if possible preprepared initial defensive positions have been selected by NATO Center Region forces. However, the very composition of the defending forces — mostly highly mobile armored or mechanized divisions — suggests that a maneuver defense is contemplated if these forward positions are pierced.

On the other hand, the thin linear defense without much depth imposed by the limited number of costly heavy divisions available to man it is one of the NATO Center Region's greatest weaknesses. It also ties down in a less than optimal role precisely those mobile forces that are best suited to be mobile reserves for counterattack or counteroffensive. It can be saturated in time by successive Soviet attacks; once broken through, a coherent defense would be hard to reconstitute because added ground forces are lacking.

Yet this defense could be greatly thickened up by a series of relatively cost-effective defensive measures that would hardly be beyond NATO's grasp. One key measure should be to capitalize on the fact that the increasing urbanization of Central Europe is already creating a form of urban barrier system that is a serious obstacle to large-scale armored maneuver. So too is Germany's large proportion of forested or mountain terrain. If Bonn would reconsider its long-standing political reluctance to fortify the inner-German border, this existing "barrier" could be supplemented by a classic economy of force measure, light fortification on key avenues of advance, and obstacles in the remaining gaps. The model here is fortifications of the Siegfried Line type rather than like the Maginot Line.[9]

This forest and urban barrier complex, bolstered by obstacles and some fortification, is well suited to manning largely by additional reservist infantry units, which could be formed from the large pool of European conscripts.[10] Another fifteen to twenty-four reserve infantry brigade equivalents — equipped at most with "wheels," not "tracks," and defending covered terrain — would do much to provide a thicker defense in depth to absorb any Soviet blitzkrieg penetration.[11] They would also help to free up the costly NATO armored divisions for their optimal counterattack role.

Nor need NATO be mesmerized by misleading equipment counts such as the three-to-one superiority of the Warsaw Pact over NATO in numbers of tanks. The equation is much more complex. Aside from antitank obstacles, proliferation of ground- and helicopter-launched antitank missiles, mines delivered dynamically on the battlefield by artillery and rocket launchers, air- or missile-delivered armor-piercing submunitions, and the like would all help equalize the balance.

Advances in nonnuclear technology further offer great promise of dislocating the Soviet follow-on forces that otherwise might overwhelm any NATO defense that held firm against initial attack. Longer-range delivery systems armed with area submunitions are already in development, and the problems of long-range target identification and prompt response seem amenable to solution. Similar air- or mis- sile- delivery vehicles for prompt attack on Soviet air bases could also cripple the tactical air offensive planned in support of a Warsaw Pact attack.[12] These would help greatly to achieve defense in depth by extending it forward of the line of battle. But they will be very costly indeed in numbers large enough to make a difference. Nor are they a panacea for all of NATO's many deficiencies. For example, what price dislocating the second echelon if the first echelon breaks through?

Rapid reinforcement of the crucial Center Region offers another crucial means of thickening up the defense, and adding powerful counterattack forces. Hence, using propositioned equipment and flying over the troops and planes, the United States plans to double its ground forces and triple its air forces in Europe in less than two weeks, provided that the Europeans contribute depots and other facilities, air bases, some of the air- and sea lift, and other essential host nation support. This "transatlantic bargain" is so favorable to Europe that it is hard to grasp why some allies still cavil over funding the modest add- ons required. Even a stout initial defense of NATO Europe would not be feasible without the enormous contribution which such U.S. reinforcements provide.

On the other hand, our NATO allies must recognize that this is the practical limit of U.S. rapid-reinforcement capability, given the wide range of other important contingency needs that the United States must meet on behalf of the Free World — not least in the Persian Gulf. In fact, Washington ought to announce a ceiling on its initial NATO ground force contribution at the ten-division level in order both to satisfy the Congress that our NATO commitment is not open-ended and to bring home to our allies that they must assume the rest of the ground force burden.

The Role of France

Under these circumstances, "French forces . . . would also have to play a substantially greater role to increase the credibility of a con- ventional defense."[13] In fact, *whether NATO can achieve a credible nonnuclear initial defense posture in the crucial Center Region may well depend on the key role played by France.* With a military budget now comparable to that of Germany, France could surely provide forces as strong as those of the Bundeswehr, were it not that so much of her investment budget goes for her independent nuclear force. In addition, it would be difficult to sustain for long the reinforcements the United States plans to send to NATO without additional lines of communication across France.[14] Such practical French cooperation with her allies, beyond what has already discreetly taken place, is far more important than whether France formally rejoins NATO's military wing.[15] Hence it is to be hoped that France will examine whether the declining credibility of the U.S. nuclear umbrella does not force a reexamination of a policy that no longer meets her own security needs — and is certainly hard on her allies.

It is equally urgent that France's allies think through what inducements might suffice to secure such a vital contribution. For example, the United States could assist in the modernization of French nuclear forces, as we already do for Britain, provided that France strengthened her conventional forces. This would probably also require reaching some kind of joint understanding over the use of French, U.S., and U.K. nuclear forces in Europe, through which France would gain in reassurance (for example, a finger on the U.S. trigger) whatever she gave up in independence.

Much more also needs to be done to provide effective air defense of NATO Europe, something that tended to be neglected when Soviet tactical air forces were designed mostly to defend their own airspace. But now their offensive configuration poses a major threat, not least to the NATO air support of its own ground forces on which NATO has so long relied as a major equalizer. Modernization of NATO's NIKE/HAWK "belt" with more modern systems, point defenses, shelters, and the like will be cumulatively very expensive, but indispensable. France too should pay her share. Moreover, if France could make available beforehand existing airfields in Northeast France, the dispersal (hence survivability) of NATO air forces would be significantly enhanced.

The foregoing brief survey is mainly designed to suggest that a high-confidence nonnuclear defense of the crucial NATO center is indeed feasible at acceptable cost. It is by no means a lost cause. NATO has the building blocks for an affordable conventional deterrent capability to complement nuclear deterrence, if the alliance can rise above the formidable political and economic obstacles entailed, not least to France's assumption of its indispensable role.

Protecting Persian Gulf Oil

Credible deterrence/defense against a Soviet military threat to Persian Gulf oil is a much trickier proposition. Emergence of a power vacuum in the Persian Gulf confronts Washington with an additional important threat to Free World interests, which only U.S. force projection could hope to deal with but which further stretches already thin American capabilities. Moreover, the United States would be at considerable geographic disadvantage in having to deploy to this remote area, almost devoid of U.S. bases or U.S. forces, except at sea. As already noted, sea-based forces could not hope to successfully defend the oil fields themselves, which are well up within the shallow Persian Gulf.

These adverse factors have led to criticism of the Carter Doctrine as an empty threat. To put teeth into it, the United States is developing a rapid deployment force (now called Central Command) of sufficient size and deployability to offer prospects of actually holding north of the oil fields themselves. However, many critics also write this off as being neither rapidly deployable nor a new force, merely a redesignation of existing forces already earmarked for yet other missions and usable only at their expense. The critics also point to the lack of available bases or even reliable allies in the region, in strong contrast to Europe or Northeast Asia.[16] There is some validity to all these criticisms.

Posturing to defend the oil fields is admittedly a high-risk strategy. But the real issue is whether the risks entailed are justified by the vital Free World need for unimpeded access to the oil. Moreover, the critics understate the crucial importance of deterrence. Even if the United States could impose only a thin "tripwire" of light forces between the Soviets and the Gulf oil fields, this would still give the Soviets pause. After all, it would mean a direct U.S.-Soviet shooting confrontation of a sort that has not occurred since 1919, with consequent risk of wider war. Moreover, this is not our only option.

If deterrence failed, the key problem would be to deploy initial U.S. forces fast enough and then to build up and sustain them. The United States has had long experience with overseas force projection (we deployed around half a million men to Korea and then Vietnam, the latter about as far away as the Persian Gulf). The crucial element would be rapidity of precautionary or if necessary preemptive deployment. To this end, the United States is prepositioning equipment on ships in the Indian Ocean at Diego Garcia, stretching its C-141s, buying 50 improved C-5Bs and many more KC-10 tankers, and converting eight SL-7 fast roll- on/roll-off ships to carry heavy equipment — all of which will greatly enhance rapid response.

The other key to rapid response — base facilities in the area — poses yet more difficult problems. Since the nations of the Persian Gulf-Indian Ocean region are politically reluctant to align themselves formally with the United States, we have shifted to a concept of helping them to develop their own "facilities" which they could make available in an emergency. As an alternative, Saudi Arabia, so rich it doesn't need U.S. subsidies, is nonetheless building an elaborate base structure which it could make available in event it saw an imminent threat. We are counting on the Gulf nations and other important ones like Egypt, Sudan, Somalia, and Kenya to recognize their own security interests if they come under threat. Let's remember that, in the last analysis, whatever peacetime base or transit arrangements have been made in any theater, U.S. forces will be able to use them only if the host nation sees it in its interest to allow such use.

Critics also allege that in posturing to cope with a Soviet military thrust, we are focusing on the least likely threat. Much more likely are internal coups in the volatile Middle East or Africa, which could bring hostile regimes to power as happened in Iran. True, other means than U.S. military intervention will have to be primarily relied on in this event. However, deterring *Soviet* intervention in such crises - in effect holding the ring — is also essential to preserving U.S. interests. It too requires a credible intervention capability.

It is also quite true that posturing existing forces for Persian Gulf or other Third World contingencies could lead to their being unavailable for NATO or Northeast Asian contingencies. However, resource constraints have always compelled us to design our forces as "general-purpose forces" alternatively available for different contingencies. Since only the United States has the costly force-projection capabilities to deal with remote contingencies like the Persian Gulf, we put NATO and Japan on notice as early as 1980 that this would necessarily entail significant diversions of forces otherwise available for NATO and Pacific defense.

The most cost-effective means of dealing with such diversions is a more "rational division of labor," as advocated in 1980 by then Chancellor Helmut Schmidt. Since European and Japanese force-projection capabilities are at best limited (only France and Great Britain have even modest capabilities), it makes military sense for U.S. forces to take on the oil-protection mission, while the other affected allies concentrate on shoring up their own home defense capabilities, thus compensating for any U.S. diversions. This is a strategic fact of life, which makes it all the more regrettable that few of our allies have yet done more than accept it in principle.

Rethinking Pacific Defense

The rapid economic growth of Japan, and the emerging parallelism of strategic interest between China, Japan, and the United States, make East Asia the theater offering the most interesting new strategic opportunities. The dominant strategic interest of all three nations focuses on deterring Soviet expansionism. This common interest also creates a potential two-front threat-in-being that Moscow cannot ignore. Already over one-quarter of Soviet conventional forces are tied down opposite China and Japan, even though neither nation has much offensive capability for threatening Soviet territory. Offering Western technology, financed directly or indirectly by Japanese loans, to strengthen China's defensive capabilities would be a classic strategic option, historically employed to deter a strong opponent by confronting it with risks of a two-front conflict.

But could the United States realistically expect to convert this potential threat-in-being into actual offensive capability in event of U.S.-Soviet or NATO-Warsaw Pact war? No. We must assume that in all likelihood, Japan and China would see advantage in remaining neutral, at least initially, in the event of a U.S.-Soviet clash (as might Vietnam). With little offensive capability, they would stand to lose far more than they gained by becoming belligerents. Without offensive help from China and Japan, it is hard to see what the United States could achieve militarily in the Pacific theater against the U.S.S.R. Bottling up the Soviet Pacific fleet (difficult without active Japanese cooperation) and a few carrier strikes at Soviet naval bases could not seriously damage Soviet war potential. In short, likely Japanese and Chinese neutrality severely limits the useful strategic options open to the Pentagon in the Pacific theater.

As suggested earlier, it is equally hard to see what the U.S.S.R. would gain from attacking Japan and China until it had achieved its logical wartime objectives in Western Europe and perhaps the Persian Gulf. This would involve a major diversion of forces and resources, whereas if Moscow only waited until it had defeated NATO and/or occupied the Persian Gulf oil fields, it would have enormous coercive power over Beijing and Tokyo. On this score, it is worth repeating how Japan and the U.S.S.R. stayed neutral vis-a-vis each other throughout most of World War II.

We need to rethink our Pacific strategy in the light of such real life prospects. If it is desirable to reduce the likelihood that the United States might be confronted with a simultaneous "three-front" war, the most promising prospects lie in East Asia.[17] In any case we cannot assume that we can drag China and Japan into a Pacific war regardless of their preferences. If we cannot, what important gains can we expect to achieve?

Dealing with Third World Contingencies

At a time when nuclear stalemate erodes the credibility of extended deterrence, yet conventional U.S. forces are stretched thin, we also need to rethink our strategy for dealing with lesser Third World contingencies. True, these are precisely the kinds of contingencies most likely to occur in the future, but the United States no longer enjoys the nuclear superiority that enabled it to divert vast resources to limited wars like Korea and Vietnam while still being confident that they would not escalate to wider war. Now such diversions could far more dangerously degrade our global deterrent position, especially if they grew beyond what was initially contemplated — as history shows they tend to do. In short, we have to be more prudent about Third World use of military force on a scale that would seriously interfere with other higher priority U.S. commitments. We need an economy-of-force policy from the strategic point of view.

This is not to suggest that the United States must forego such interventions regardless of the cost to U.S. interests, but that we must be much more chary of the possible direct and indirect costs. To the extent feasible, anticipatory diplomacy, military and economic assistance (including advisors), and even precautionary deployments or covert actions loom larger as preferred options from this viewpoint. Should compelling reasons lead to overt military intervention, then we should to the fullest extent feasible use flexible sea and air power instead of ground force commitments that tend to escalate as we get bogged down. Carriers do have their uses. So do quarantine, blockade, and even mining, should the necessity arise.

Although this book addresses mainly strategic rather than force-posture issues, our strategy would be better served by greater flexibility in our ground/air posture, tailoring it better to foreseen Third World contingencies, not least in the Persian Gulf. For example, lighter, more rapidly deployable forces — able to operate on an austere basis from austere bases — seem generally better suited to quick response needs than our current capabilities. Greater strategic mobility via fast sea lift and airlift is another sine qua non. All this would also help enhance deterrence.

Let me briefly argue the virtues of an economy-of-force strategy aimed at deterrence by taking Latin America as a case in point. Although some argue that it is not of vital strategic interest, few deny that we would feel compelled to respond vigorously to an overt Soviet (or Cuban surrogate) intervention with military forces, or even to a Soviet attempt to acquire a military base. To forestall the need for a military response by alert anticipatory efforts is obviously our best course. Such a deterrent policy requires not only a clearer definition of U.S. interests but also combined political, economic, and military aid measures, building on two facts of life. First, Latin America is in our own strategic backyard, making it hard for the U.S.S.R. to project force at such a distance (just as Eurasia's location in the Soviet backyard makes it hard for us to respond militarily to such actions as Soviet occupation of Afghanistan). Second, and by the same token, our Latin American colleagues fear nearby U.S. hegemonism more than they fear the remote Soviets. Moreover, the simple presence of hostile regimes like Cuba is not a satisfactory *casus belli*; it becomes one only if Cuba is used by the U.S.S.R. to create a major military threat to U.S. interests.

Among the deterrent measures whose utility has not always been adequately understood by Congress is the provision of military aid and advisors as an economy-of-force measure. Largely because of our bitter Vietnam experience, many members of Congress feel that this increases the likelihood that we will end up sending troops. On the contrary, experience suggests that this can forestall the risks of deeper entanglement. As McGeorge Bundy aptly put it with respect to El Salvador, "the notion that a reforming reinforcement of our military advisory process . . . would carry us across some Rubicon to overcommitment is pure fancy, an illustration of our fondness for learning the wrong lessons from our military history."[18]

Given the mismatch between our strategy and our resources, we also should carefully examine more radical alternatives for which there is a compelling strategic case. Most U.S. interventions since 1945 have been to contest with perceived communist surrogates, real or potential. Our current concern with Central America proceeds less from our fear of revolutionary change in small countries than from our fear lest Soviet/Cuban exploitation of local revolution create a growing strategic problem in our own backyard. Hence our strategic interest lies in decoupling political change from these strategic consequences. We can more readily live with radical states provided that they are not used against us by the other superpower. Yugoslavia and China are two cases in point.

Coping with Soviet Surrogates in Event of a U.S.-Soviet Clash

Above all, we want to prevent Soviet exploitation of such states as Cuba, Ethiopia, Vietnam, South Yemen, or Angola in wartime, when having to cope with them militarily could involve major diversions of forces from other pressing needs. If we are really less concerned that they have extreme leftist regimes than that these might be used against us, then the elements of a trade-off arise. Such countries as Cuba, South Yemen, Vietnam, Ethiopia, and Angola are well aware of their vulnerability to offshore U.S. power, given Soviet difficulty in supporting such remote areas in a crunch.

Take Cuba as an example and look at the case of what to do about it in event of a U.S.-Soviet clash or imminent threat of one. Obviously Cuba's geographic locale athwart the Caribbean shipping routes would enable offensively configured Cuban or Soviet forces using Cuban bases to threaten important trade and military traffic. Cuban bases could even be used to mount a serious threat against the continental United States (which led to the 1962 missile crisis). On the other hand, Cuba is highly vulnerable to U.S. retaliation from surrounding bases. It is also totally dependent on sugar and tobacco exports for its livelihood, and highly dependent not only on Soviet economic aid but also on imported fuel.[19] All this offers possibilities for a carrot-and- stick approach.

Washington could make clear to Havana that if it held aloof from a U.S.-Soviet conflict and did not allow Soviet use of its bases, we would not only respect Cuba's territorial integrity but guarantee both its sugar/tobacco exports and its imports of fuel. We could further undertake not to try to overthrow the Castro regime after the war was over, and even to return Guantanamo. On the other hand, we could threaten that if Cuba did allow Soviet offensive use of its facilities, the United States would seal off Cuba by blockade and mining, totally preventing any imports or exports, interdict all Cuban shipping and air traffic; raid and harass Cuban territory; and even possibly invade. If Castro is halfway rational, it would not be difficult for him to figure out that even if the U.S.S.R. eventually won a U.S.-Soviet conflict, Cuba would be a casualty.

A similar scenario of carrots and sticks could be devised for other pro-Soviet countries beyond the effective range of Soviet wartime power. They too would be likely to suffer disproportionately if they allowed the U.S.S.R. to use them in wartime. Moreover, most of these nations appear to be nationalist first and pro-Soviet only second.

Who can tell if such strategic options would actually work? Serious problems of credibility, even of communication, would arise. Timing of any U.S. offer would be crucial, and we would have to hedge against the possibility of failure. Nonetheless, the strategic advantages are such that these options should be carefully studied in advance and politico-military contingency plans prepared if they seem promising. If the carrot-and-stick approach failed, it would have cost us nothing and left us no worse off then if we had never tried it. Nor will it escape the reader that if such options are strategically desirable in wartime, this has obvious implications for our peacetime policy as well. Here is another example of the all too frequent disconnect between our peacetime policy and our wartime strategic needs.

To sum up, there are a number of strategic options, including economy-of-force measures, that could help reduce the strategic impact of the mismatch between our strategy and our resources. This cursory review is designed to help stimulate their examination. Some of these options would also be much more congenial than present U.S. policies to most of the allies on whom our global deterrent/

defense posture is so dependent. On the other hand, they all entail political complications and other practical obstacles that have been glossed over here. For example, a stalwart initial nonnuclear defense of Europe would entail significant increases in allied contributions to the common defense, which the next chapter will discuss.

NOTES

1.   Stansfield Turner and George Thibault, "Preparing for the Unexpected: The Need for a New Military Strategy," *Foreign Affairs*, Fall 1982, pp. 123/124, call it "an argument against change" that "typifies the resistance of Atlanticists." Michael Vlahos terms it the "orthodox school of national security posture" in "Maritime Strategy vs. Continental Commitment?," *Orbis*, Foreign Policy Research Institute, Fall 1982, p. 583.

2.   See for example Karl Kaiser, George Leber, Alois Mertes, Franz-Joseph Schulze, "A German Response to No First Use," *Foreign Affairs*, Summer 1982, pp. 1157/1170.

3.   James Schlesinger, "The Handwriting On the Wall May Be a Forgery," *Armed Forces Journal*, March 1982, p. 28.

4.   Josef Joffe, "Europe and America: The Politics of Resentment (Cont'd)," *Foreign Affairs*, volume on *America and the World 1982*, June 1983, p. 569.

5.   As Kaufmann shows in *Planning Conventional Forces 1950-1980* (Washington, D.C.: Brookings Institution, 1982), this same approach has been used before to reconcile force sizing with strategic needs.

6.   Incidentally, these arguments show why there is no point in the United States preparing for a "protracted" conventional war in Europe unless we can get our allies to do the same.

7.   Francois de Rose, "NATO's Perils and Opportunities," *Strategic Review*, Fall 1983, p. 22.

8.   Kaufmann suggests a number of options for strengthening NATO Center Region conventional defense, generally similar to those of the author, in *Alliance Security: NATO and the No-First-Use Question* (Washington, D.C.: Brookings Institution, 1983), pp. 43/90.

9.   Kaufmann offers some interesting calculations on the defensive impact of barriers (ibid., pp. 65/72).

10.   Since Germany and other Center Region allies face a declining manpower pool, conscription cannot produce much larger active forces. However, more use of reservists would help circumvent this problem. It would also get around, in the key German case, the WEU prohibitions on active German forces larger than 500,000, although these prohibitions themselves — designed to meet a quite different problem 35 years ago — should be reviewed.

11.   A provocative analysis by General Franz Uhle-Wettler, a serving Bundeswehr officer (Gefechtsfeld MittelEuropa: Munich, Bernard & Graffe Verlag, 1980) makes a convincing case for light infantry fighting in such covered terrain.

12.   Development of these systems has been strongly endorsed by General Rogers, German Defense Minister Manfred Woerner, Secretary of Defense Weinberger, and Senator Sam Nunn. A strong case for them is made in the Report of the European Security Study, *Strengthening Conventional Deterrence in Europe* by a group of 26 American and European experts (Macmillan Press, London, 1983).

13.   Harold Brown, *Thinking about National Security*, p. 103.

14.   For the best recent analysis, see David S. Yost, "France's Potential Contributions In Conventional Contingencies in Central Europe," paper prepared for European-American Institute Workshop, Washington, D.C., October 29/30, 1983.

15.   De Rose takes a similar view. He too argues that "a needed push from France is crucial to greater NATO conventional strength." See "NATO's Perils," p. 26.

16.   See Jeffrey Record and Robert J. Hanks, *U.S. Strategy at the Crossroads* (Washington, D.C.: Institute for Foreign Policy Analysis, July 1982), pp. 17/28, for a scathing critique. The trouble is that Record nowhere discusses the fatal flaws in the "sea borne force projection" alternative he advocates (p. 29).

17.   In *Rethinking Defense and Conventional Forces* (Washington, D.C.: Center for National Policy, 1983), p. 49, I suggested developing "two armies" - one of heavy forces for reinforcing NATO or Korea, and another of lighter, more mobile forces primarily for other contingencies.

18.   Letter to *Foreign Affairs*, Fall 1983, p. 203.

19.   See Mark N. Katz, "The Soviet-Cuban Connection," *International Security*, Summer 1983, pp. 88/112.

CHAPTER 9

Generating Adequate Coalition Resources

While the foregoing strategic options would make a Eurasia-oriented conventional strategy easier to execute, the basic question is less one of strategic desirability than of whether the alliance will be able to generate adequate resources to underwrite it. This in turn is less a question of economic feasibility than of political will. Can the Western allies develop sufficient sense of common purpose, willingness to sacrifice, and collective determination for this purpose? For this to happen, the United States and its allies must face up to two facts of life — that conventional capabilities are far more expensive than nuclear, and that a United States declining in relative economic power can no longer carry so much of the mutual defense burden (see Chapter 4).

Even though our allies already bear a larger part of the mutual burden than most Americans seem to recognize, those who criticize the continental or coalition school by attacking them for not contributing enough still have a point. Whether or not "equality of effort" has existed until recently, today the United States is the only ally that has recognized the need for a greatly enhanced conventional effort and is making major added contributions to it. This also creates a problem of U.S. perceptions to which our allies must respond in their own self- interest.

Nor is it enough to rely on the far superior collective economic strength of the Free World alliance. The Soviet command economy has done a much better job of translating inferior economic strength into ready military power, in contrast to the traditional reluctance of pluralistic democracies to spend adequately on defense. This leads inevitably to the conclusion that a coalition rather than a unilateralist approach is the only viable way to achieve credible deterrence/defense at a cost politically acceptable to free societies. Thus the real issue that the United States must confront is whether a vigorous coalition effort to rectify deficiencies offers sufficient promise to be preferable to abandoning the effort in favor of a primarily maritime strategy.

A Change in Mindset is Required

The first requisite is a basic change in mind-set — a broad recognition on the part of the United States as well as its allies that a genuine coalition effort is imperative if we are to generate the needed resources and use them more efficiently. Up to this point America's allies, used to living under our nuclear umbrella, have not yet absorbed the extent to which the advent of nuclear stalemate has eroded its credibility. Nor have allied publics and parliaments yet perceived themselves as threatened to the point where a consensus for increased defense spending would emerge. Although large segments of allied publics are becoming more and more disenchanted with the nuclear arms race, this has not yet extended to a new interest in conventional defense, as has already occurred in the United States. Moreover, many Europeans still fear a replay of World War II's destructiveness, that they think would over time create casualties and war damage comparable to that from a nuclear exchange.[1] Indeed, U.S. talk about protracted conventional conflict tends to feed these fears, which suggests that NATO must reemphasize that deterrence, conventional as well as nuclear, remains its primary aim.

Nonetheless, a new incentive is coming into play. *To the extent that our allies come to perceive as we do the need for a strong conventional deterrent to offset the eroding credibility of the U.S. nuclear umbrella at a time of nuclear stalemate, it should be possible to fashion a stronger consensus to this end.* At the least the United States should put this proposition to its allies, exerting the leadership role those allies await.

Another deeply held view is that high-confidence conventional deterrence/defense is so expensive as to be impractical. Yet Chapter 8 has suggested that, in the case of NATO at any rate, many options are available to strengthen it at only moderate cost. The fact is that NATO is already within shooting distance of sufficient outlays to produce a robust initial conventional defense — one capable of deterring or halting an initial Warsaw Pact blitzkrieg, especially in the crucial Center Region. If we accept SACEUR's own estimate that 4 percent real growth for the next six years would achieve his force goals, then NATO has already spent and is spending the great bulk of what is regarded by its own military authorities as the minimum essential for such a defense.

For example, increasing the sustainability of existing forces (mainly a matter of buying more consumables like ammunition) is generally much cheaper than buying the forces themselves. Hence many Americans find it hard to understand why allies who have already bought most of the requisites now balk at the rather modest add-ons (averaging far less than 1 percent of GNP per annum) necessary to ensure that the rest of their investment is worthwhile.

Moreover, the solution lies as much in more efficient outputs as in greater inputs. It bears repeating that NATO still spends more collectively on defense than does the Warsaw Pact. Adding in other allies like Japan, we are far ahead. But we spend much less efficiently, and our high manpower costs reduce what goes to investment. Therefore, higher spending is not the only way in which additional capabilities might be generated. More sensible burden-sharing and more efficient collective use of the resources available would also pay high dividends.[2] Indeed, the more rationally collective burdens are shared and the more efficiently defense outlays are utilized, the less the needed increase in defense spending. As Secretary Brown used to tell his fellow NATO defense ministers, if NATO could increase defense spending by 3 percent in real terms but also achieve a 3 percent improvement in the efficiency with which we utilized defense outlays, the alliance could in time readily cope with the Warsaw Pact threat.

Here too a change in mind-set is required. Politicians, soldiers, bureaucrats, and industrialists must be brought to realize that only by organizing the common defense on the basis of "balanced collective rather than individually balanced national forces" can we achieve the goal at a cost that is politically tolerable in peacetime. This is as true of the nonnuclear defense of Northeast Asia or Persian Gulf oil as it is of Europe. Yet this flies in the face of that still potent nationalism that understandably dominates the outlook of nation-states. Our parliaments and national security establishments are used to thinking in terms of national, not multinational, defense. The "sin of unilateralism" still reigns supreme. This deeply ingrained pattern of thought poses an enormous obstacle to the evolution of an adequate yet affordable collective defense. Sheer bureaucratic inertia — the ingrained tendency of institutions to keep thinking and acting according to well-established habits — will also be hard to overcome. Yet these patterns are changing, albeit at a glacial pace. This chapter suggests a number of desirable ways in which the pace can be speeded up in our own security interests.

Need for U.S. Leadership and Inducements

But it is only stating the obvious that a *sine qua non* must be vigorous, sustained, and generous U.S. leadership. Whether wisely or not, our allies still at bottom rely on Washington to provide the lead. In this case leadership must mean more than rhetoric and persuasion mixed with complaint and pressure. To be credible, it must entail generous concrete actions to show that we ourselves are in fact moving toward a coalition strategy and posture.

Experience shows that *getting the allies to join us in this enterprise requires stressing incentives rather than threats.* It should not be beyond the grasp of U.S. statesmanship to create inducements to greater cooperation by making offers so attractive to our allies that they are too good to refuse. To overcome acute allied suspicions that greater alliance cooperation, standardization, and the like are mainly rationales for greater U.S. sales at the expense of European industry, these U.S. measures must include accepting more of a two-way street in reciprocal defense procurement. For example, if we really want our allies to join us in funding "deep strike" systems to dislocate the follow-on echelons of an attacker (see Chapter 7), it will be imperative to allow allied industry to participate fully in their development and production.[3] Any added costs to the United States would be marginal compared to the security gains. In fact, there would probably be no net add-on costs over time, because it is hard to imagine that the United States would end up with less than its fair share of expanded military trade or of any defense budget savings from other sharing arrangements.

Similarly, we should offer to share technology as an inducement, with due regard to the risk that it might leak to the U.S.S.R. Our present policy stress on preventing such leakage, even at the expense of sharing needed technology with our allies, needs to be replaced by a more balanced case-by-case approach which weighs the potential gains against the losses. In fact, we should be able to use technology sharing and greater access to the U.S. market as inducements to our allies to impose tighter controls on technology leakage.

Increasing Defense Outlays

Nonetheless, given the inevitably slow process of improving the efficiency of collective defense outputs, some increase in the present modest rate of allied defense inputs is essential to meet already established military goals. NonU.S. NATO defense outlays increased an average 2.7 percent in 1980 and 1981, but were only half that in 1982 and are probably still less in 1983. As a percentage of gross domestic product total allied outlays averaged a modest 3.6 percent.

Canada and Denmark made the least effort — 1.9 percent and 2.5 percent of GDP respectively, in 1981.[4]

With most of NATO Europe starting to recover from recession, it is hard to make an economic argument that it could not readily sustain spending at least 4 percent of GNP on defense. Assuming gradual economic recovery, this would permit substantial growth in defense outlays over the rest of the decade. The 3 percent real growth formula adopted in 1977 and reaffirmed by the 1978 NATO summit served a useful purpose in providing a feasible goal that helped achieve greater increases than were otherwise likely. It would have achieved even more but for the recession that soon occurred. This is presumably why SACEUR has called for a 4 percent average real increase which he contends would enable NATO's agreed 1983/1988 force goals to be met.[5] The United States should press NATO to adopt such a pledge, perhaps at another summit.

Japan, of course, is a special case. With the world's third-largest GNP and still impressive economic growth, Japan's defense outlays of less than 0.9 percent of GNP make it an obvious candidate for greater burden-sharing, especially since it is the largest single consumer of Persian Gulf oil. Moreover, Japan's commercial competitiveness is increased at the expense of its allies when it contributes so little to the common defense. Under these circumstances, Japan is becoming a favorite target of U.S. defense planners, who seek ways to get Japan to contribute more to its own defense, including that of the adjacent sea-lanes on which its livelihood depends.

What holds Japan back is primarily lack of domestic political consensus rather than economic weakness. Fitful U.S. efforts to persuade a reluctant Japan to assume more of the responsibility for its own home defense have had some impact, but clearly much more is needed. The United States should be more candid with Japan that in event of a multifront conflict, or even a major U.S.-Soviet clash elsewhere, we would not be able to make the contributions previously expected to Japan's conventional defense. If we must defend the Persian Gulf oil on which the Japanese economy is so dependent, we cannot simultaneously spare enough forces to secure Japan itself, or South Korea for that matter.[6] If, as suggested earlier, Japan would probably prefer to remain neutral in most contingencies other than a direct threat to itself, then a stronger self-defense capability would certainly strengthen its ability to do so (the Swedish model is apposite here). The Japanese government owes it to its own populace to bring these facts to their attention and to propose a more realistic defense program.

Given Japan's special political problems, its further contributions to the common defense could be partly indirect (as some are already).

Increased Japanese economic aid to threatened countries around the Soviet periphery would free these countries to spend more of their own resources on defense. For example, South Korea has had a good case in seeking concessional loans from Japan, on the ground that Korea, in spending over 6 percent of GNP on its own defense, is providing Japan a better defensive buffer zone.

More Rational Burden-Sharing

Since modest real defense spending increases averaging 3-4 percent per year are unlikely by themselves to meet conventional needs, rationalization of defense missions between the allies (often called specialization) also seems required. Indeed the present SACEUR finds "specialization" the main direction in which we should go.[7] It is essential to help overcome the wasteful overlapping of capabilities that results from most allies still seeking to maintain full-spectrum defense establishments as if they were going to fight alone.[8]

Of course, a degree of specialization already exists. The United States has long assumed the main nuclear deterrent role. The U.K. and French nuclear forces also contribute. Similarly the United States has taken on the principal blue-water naval role, though Britain contributes extensively to control of the crucial North Atlantic sea lanes. But several continental countries — particularly Germany, Holland, and Belgium — maintain for essentially political and traditional reasons naval forces larger than their primary need to cope with land or air threat from the East would justify from a purely military viewpoint. There is an important potential for rationalization here.

In the Mediterranean, not only the U.S. Sixth Fleet but powerful French and Italian and Spanish plus the Greek and Turkish navies together outweigh the Soviet presence. As carriers become increasingly vulnerable in these narrow seas, NATO should develop better plans for replacing them by largely allied missiles, land-based air, submarines, and light attack craft. True, the U.S. Sixth Fleet also performs nonNATO missions involving the Levant and North Africa. However, task forces could quickly be deployed from the United States to cope with crises in these areas.

Greater specialization in land and tactical air forces is feasible too. Air defense, for which NATO is trying to develop an integrated program, is one obvious field. The United States is already exploring options whereby it would provide the new PATRIOT system for air defense of the Center Region if the allies would provide and man point-defense SAM systems, and contribute other offsets.

Perhaps the most far-reaching potential for what Helmut Schmidt called a "rational division of labor" entails the United States assuming primary responsibility for the oil-rich Persian Gulf region while our European allies replace the U.S. forces diverted to this mission by contributing more in NATO (see Chapter 8). Regardless of whether Europe needs Persian Gulf oil more than the United States does — a dubious proposition given the global economic impact of another oil shortage — the fact of the matter is that only the United States has the remote-area force-projection capability to protect it.[9] Redistribution of naval responsibilities in the Mediterranean would be one important step in this direction.

Turning to the Far East, a greater specialization of missions for defending Japan and Korea is also overdue. The only U.S. division left in Korea is there primarily for reassurance and deterrence reasons, not because our one division adds all that much to nineteen U.S.-equipped Korean active divisions and twelve reserve divisions. Given the urgent needs for greater U.S. ground forces in other contingencies such as the Persian Gulf, the U.S. contribution to deterrence/defense of South Korea and Japan must be primarily via air and naval power. Already the U.S. Marine division on Okinawa, the nearest available U.S. ground reinforcement for Korea, has a high-priority mission for Persian Gulf deployment. If Japan can eventually be prevailed on to assume greater responsibility for helping in Korea's defense, the U.S. Army division in Korea might be redeployed.

As for defense of Japan itself, the United States should make clear for similar reasons that no U.S. ground forces can be made available for this purpose. Indeed, we should insist that Japan gradually take on full responsibility for its own ground and air defense, as well as for protecting its sea-lanes out to 1,000 mile range, while the United States contributes the nuclear deterrent, long range conventional striking power, and the like. Beyond this, if Korea needs ground reinforcement, this should logically become over time a Japanese responsibility as well, since Korea serves as a critical buffer zone for Japan. In sum, because U.S. forces are spread so thin, we must bring our Japanese and ROK allies to recognize that, whatever their mutual antipathies, they themselves will have to provide most of the ground forces for their own mutual defense.

Greater Armaments Cooperation

Another broad field in which greater efficiency in coalition outputs is essential is that of armaments cooperation. Although NATO has already achieved a degree of peacetime defense cooperation (see Chapter 3), what has already been accomplished only scratches the surface. It does, however, suggest the enormous possibilities if only national particularism and bureaucratic inertia could be overcome.

Greater armaments cooperation would not save a lot of money, at least in the early phases, unless it entailed a greater degree of specialization. This was the intent of the weapons-family approach discussed in Chapter 3. Multinational R&D or production projects tend to cost more than if they were handled entirely by one nation, but this is not the point. Given the escalating costs of today's technology, only via multinational projects can most nations put together enough capital to fund them in the first place.[10] Moreover, the cost to each nation is much less than if it tried to go it alone. This rationale lay behind such European projects as Tornado, Jaguar, Hot, Milan, and the like.

In any case, there are many other techniques by which armaments cooperation can be achieved, each with different impacts. Straight cross-licensing of new designs for national production at least saves large R&D outlays. Using the weapons-family approach for dividing up R&D is another. Single-country production for sale to other allies, often with offsets, is an efficient technique, as is coproduction on the F-16 model, where the United States is prime producer but several NATO purchasers make components, assemble aircraft, and also receive some other offsets. Creating new multinational corporations like Panavia or Euromissile has its advantages. Industrial teaming on both sides of the Atlantic is growing, as industry again proves more farsighted than government in seeing that cooperation is the coming thing.

One spinoff from greater cooperation would be greater interoperability and standardization — so important for coalition war. Without the same calibers of ammunition, interchangeable fuels, compatible communications, and the like, it is hard to see how NATO's national contingents can operate optimally together. Regrettably, the only coalition that is today enforcing standardization across the board is the Soviet-dominated Warsaw Pact, where the Soviets impose it by fiat. Clearly, the NATO military authorities must place greater emphasis on their need for compatible equipment, tactics, and procedures than they have to date. This has been one of their greatest failings. Unless they insist on common standards, why should the politicians respond? Moreover, it is the military who will be the operational beneficiaries.

But it is governments and politicians who will have to provide the real impetus. They must insist on integrated multinational programs, as were mandated in the Long-Term Defense Program of 1978. Its initiatives for an integrated air defense, a common C(3) program, standardized electronic warfare systems, and common munitions stockpiles deserve to be reinvigorated. As called for in the recent Roth-Glenn-Nunn amendment passed almost unanimously by the U.S. Senate, we must pool alliance industrial resources more efficiently to avoid wasteful duplication and achieve economies of scale. Standardization or at least interoperability also must be insisted on by governments, lest their lack contribute to disaster on the battlefield against the much more homogeneous Warsaw Pact forces.

It is quite possible to rationalize armaments development and production while still allowing *juste retour*, the equitable dividing up of jobs, profits, and export opportunities most countries insist on. What is needed to this end is large-scale trade-offs and imaginative use of the whole panoply of techniques for industrial cooperation. Here is where more of a two-way street comes in.

An equally strong case can be made that the European Economic Community should take the lead in rationalizing European defense industry, as it is doing with other industrial sectors, because it is the only agency with the clout and staff to do so.[11] Integrating European defense industry would also make Europe a more equal partner in negotiating with U.S. industry.

Greater Host Nation Support

Another important way to limit the drain on constrained U.S. resources is for our allies to provide greater logistic and other support in wartime and even peacetime to U.S. forces either maintained overseas or prepared to deploy rapidly forward in a crisis. Our forces simply cannot rapidly deploy in the strength so vital to deterrence in the 1980s if they must also carry along the lavish level of logistic and other support traditional to the American style of war. Nor does it make sense to use constrained U.S. resources to provide such support whenever the allies or friends to whose succor we are deploying can readily provide it for us. In fact, they can supply much of it simply by mobilizing assets from their civil economies, as they already plan to do to support their own forces in wartime. They could readily do so for us as well, at essentially no peacetime cost to them or us. Therefore, the United States *should pursue a global policy of relying on wartime host nation support wherever feasible,,* and be prepared to pay for its wartime costs (on the model of Lend Lease).

We should also as a firm policy ask our allies to plan on mobilizing more reservists to support U.S. forces on their soil in wartime. A case in point is the generous German agreement to mobilize 93,000 reservists for this purpose, at one-tenth to one-quarter the cost of maintaining this number in our own active or reserve forces. This is such a beneficial form of burden-sharing that it is difficult to grasp why Congress refused in FY 1983 to pay the first phase U.S. share of the modest costs entailed.

As another form of rational burden-sharing, those nations that · can afford it should be asked to pick up *all* the peacetime stationing costs of U.S. forward-deployed forces. Many of our allies already provide extensive peacetime support of this nature. Germany, Korea, and Japan, for example, each contribute over $1 billion annually in direct or indirect support payments, labor, foregone taxes, and so on. Even if they paid all add-on costs, the United States would still be spending far more than they would for the people and equipment costs of the forces themselves.[12]

Role of Security Assistance

The strategic rationale for aiding needy allies so that they can better defend themselves is as valid today as when President Truman first stated it in 1947. All succeeding administrations have found it perhaps the most effective overall technique of coalition burden-sharing employed by the United States. In most cases it has been a useful marriage of trained local manpower and U.S. munitions. From Lend Lease in World War II until the present, either grant aid or different forms of subsidized loans or sales have been employed extensively. It is of course impossible to say what deterrent impact this has had, but logic suggests that it has been substantial. Even in cases where deterrence failed, as in Korea and Vietnam, local forces equipped and supplied by the United States carried a large part of the combat burden. Aid has also served as a useful form of "rent" for important U.S. base facilities abroad.

Nor has such security assistance actually been as costly to the United States as the tens of billions of dollars officially allocated would imply. Most military aid in the late 1940s and 1950s was World War II surplus, and the practice of providing secondhand U.S. equipment to allies (incidentally facilitating the purchase of more modern equipment by U.S. forces) has continued to this day. We gradually shifted from grants to sales (usually on subsidized terms), but sales also serve as a form of export promotion. In effect, a dollar prudently spent on military aid generally pays a higher rate of return at the margin than the same dollar spent on our own forces: Not only does the ally being aided pay for its own forces but these forces are already located in the area of perceived threat.

Regrettably, a suspicious Congress often has been unwilling to provide as much military aid as successive administrations have sought, despite its being a highly effective form of burden-sharing that helps deter conflict and "increase the ability of our friends and allies to defend themselves without the commitment of U.S. combat forces."[13] During the 1950s our security assistance budget ranged from 5 to

10 percent of the defense budget, whereas today it is about 1.5 percent, although our needs are surely much greater. Moreover, as in the cases of Greece and Turkey, political considerations can come into play. Nonetheless, any sensible coalition approach entails impressing more vigorously on the Congress that it is usually cheaper to aid allies than to have to maintain even larger U.S. forces to rescue them in a crunch. Even in the case of allies who are economically able to pay for their own equipment but are politically unwilling to do so, it would make strategic sense for the United States to provide this equipment if the allies would man and maintain it. In this way America would again serve as the "arsenal of democracy" in supplying forward-deployed allied or other friendly forces.

To sum up this chapter, the need for greater conventional strength at politically acceptable cost makes more of a coalition approach imperative. Realistically feasible increases in country defense outlays are essential but unlikely to suffice. Thus more rational burden-sharing, a more sensible division of military missions, expanded armaments cooperation, and greater HNS are vital to help fill the gap. Though the obstacles are enormous, the declining credibility of nuclear deterrence and the high cost of a conventional complement may at long last provide sufficient incentive to overcome them over time. To capitalize on it, however, will take vigorous and consistent U.S. leadership not just by the Pentagon but by the White House and State Department. A vital ingredient to this end must be a set of convincing inducements to our suspicious allies.

NOTES

1. De Rose, "NATO's Perils," p. 22.

2. Secretary of Defense Weinberger makes this point compellingly in "U.S. Defense Policy," *Atlantic Community Quarterly*, Fall 1983, p. 260.

3. This point is made vigorously by De Rose in "NATO's Perils," p. 24.

4. *DoD Report on Allied Contributions to the Common Defense*, March 1983, pp. 2, 23. Rather than GNP NATO uses gross domestic product, which excludes return on overseas investment, for comparability purposes.

5. Interview with General Bernard Rogers in *Armed Forces Journal*, September 1983, pp. 72ff. Also Rogers, "The Atlantic Alliance: Prescriptions for a Difficult Decade," *Foreign Affairs*, Summer 1982, p. 1155.

6. When reviewing the first joint U.S.-Japanese plan for defense of Japan if it alone were attacked (admittedly an unlikely case), I was stunned by the huge U.S. force requirements and the very modest contributions expected of Japan itself. It was simply unreal.

7.   Rogers Interview, pp. 78, 80. The chief areas of prospective specialization he cites are new systems for attack on Soviet follow- on forces and for NATO air defense.

8.   Interview with Sir Frank Cooper in *Defense Attache*, No. 4, 1983, pp. 53/54.

9.   In a 1983 International Energy Agency (IEA) contingency exercise assuming a cutoff of Persian Gulf oil, its price rocketed to $100 per barrel. Not only would this have catastrophic global economic effects from which the U.S. economy could not be insulated; but, as an IEA member, the United States is obligated to share its oil supplies with other affected countries. "U.S. Vulnerability to Oil Cutoff Seen," *New York Times*, September 18, 1983, p. 11.

10.   Sir Frank Cooper, former permanent undersecretary of the British Ministry of Defense, criticizes Britain's high one-to-three ratio of R&D to production outlays. He says it ought to be more like one to six, so that the British forces can buy more equipment. Indeed, Britain spends a higher proportion of its defense budget on R&D than does any other NATO nation — about a third higher proportionally than the United States. Interview in *Defense Attache*, No. 4, 1983, p. 53.

11.   Surely the EEC would be more effective than EUROGROUP, or the Independent European Program Group set up specifically for this purpose. It has been a bust.

12.   See R. W. Komer, *Rethinking Defense and Conventional Forces*, (Washington, D.C.: Center for National Policy, 1983), p. 49.

13.   Speech by Judge William Clark, Center for Strategic and International Studies, Georgetown University, May 21, 1982.

CHAPTER 10
Summing Up

This effort to address looming issues of nonnuclear strategy and posture which the United States and its allies must confront in a period of nuclear stalemate is necessarily incomplete. But it will have amply served its purpose if it at least illuminates certain key issues involved in the inevitable transition from primary reliance on nuclear deterrence toward greater stress on conventional deterrence as well.

In making this transition, we must take into account the painful fact that adequate conventional forces are much more expensive than nuclear. This accentuates the "mismatch between our strategy and resources" about which our JCS complain. To the extent that resources remain constrained, as is endemic in democratic societies, we must rethink our strategy, and try harder to gear our force posture to our strategic priorities. Indeed, the essence of contemporary strategic decision making is to face up to the necessity for tough choices when we cannot do everything we want.

The Reagan administration, though seeking great increases in U.S. defense spending, has ducked this necessity for choice. Instead it has unrealistically called for an ambitious eclectic strategy of meeting a wide range of strategic requirements simultaneously, even including costly preparations for protracted conventional as well as nuclear war. In effect this has actually widened the gap between our strategy and our resources, to a point politically difficult to close.

Instead of strengthening alliance cohesion, the administration has also shown a predilection for "global unilateralism." It has focused primarily on rebuilding U.S. capabilities, rather than on concerted effort with our allies. Its policies have led to serious stresses in alliance relationships. This is all the more unfortunate because the relative decline of U.S. economic power at a time when nuclear stalemate compels greater reliance on costly conventional forces, makes us more dependent on allies than ever before. *In fact our greatest remaining strategic advantage over the U.S.S.R. is that we have many rich allies whereas it has only a handful of poor ones.*

As many predicted, these chickens have now come home to roost. Competition from domestic programs, plus the prospect that huge budget deficits might stall our economic recovery, has led Congress to cut the administration's projected defense budget increases in half. Nor have our allies, affected by the recent recession, done much to share the burden more rationally. Thus, regardless of the administration's declaratory strategy, U.S. defense budgets are again being stretched out. We cannot avoid the necessity for choice.

Hence it is crucially important that choices among competing resource allocations be illuminated by a clearer perception of their likely strategic consequences — the central purpose of this book. For these consequences will not be lost on our alert enemies or allies, who carefully analyze what we do as well as what we say. The tough choices the White House, Pentagon and Congress will have to make between various missions and between various capabilities to execute them will over time be tantamount to deciding what strategy we can actually carry out.

Indeed, this is already happening. Whatever its intentions, the administration has become locked into building a costly 600-ship Navy around fifteen carrier battle groups, which is proving so expensive (on top of a costly strategic nuclear buildup) as to crowd out other needs. The Secretary of the Navy himself boasts that, with most of the 600 ships he desires already under contract, his navy will get what it wants. But, as many in the Pentagon and on Capitol Hill are now complaining, this will be at the expense of the other services, given likely resource constraints. Investing so much in building the kind of costly carrier-heavy navy now contemplated will compromise our ability to help hold onto Europe, Japan, and Persian Gulf oil. *Thus the administration and Congress seem to be backing into a peripheral maritime strategy by default.*

The basic flaw in any maritime strategy is that, even if we swept the other superpower from the seas and pulverized all its naval bases, this would not suffice to prevent it from dominating the Eurasian landmass. Sacrificing our vital strategic interests in Europe, Northeast Asia, and the Persian Gulf would decisively shift the ultimate balance of power. True, maritime supremacy would enhance U.S. capabilities to deal with Third World contingencies, more likely to occur in the 1980s than direct attack on our vital interests. But to do so at the

expense of our primary strategic interests is to fall prey to the "likelihood fallacy" of posturing mainly to deal with the most likely
threats instead of the most serious ones. Let us also recognize how
much a maritime strategy would undermine the very network of
alliances on which a United States declining in relative economic
power must increasingly depend. It is a recipe for disaster if deterrence fails.

This is not to downplay the continued great importance of a
strong sea control navy to any U.S. strategy and posture adequate
for preserving the balance of power. "The continental and maritime
modes of operation" are "complementary not alternative strategies."[1]
As Michael Vlahos points out, "the choice in America's future strategy
is not an artificial one between 'Maritime' versus 'Coalition' defense."[2]
Obviously we need both. But the fact of the matter is that we are
seriously underfunding the latter in favor of the former. We cannot
afford to sacrifice a balanced conventional strategy and posture tar-
geted primarily on preserving vital U.S. interests in Eurasia (our
"continental commitment") in the effort to build up our ability to
cope with lesser Third World contingencies or launch carrier strikes
around the Soviet periphery which cannot decisively hurt the USSR.

The far preferable alternative would be to move further toward
a genuine coalition strategy and posture, even with the restrictions
on our own freedom of action that this might entail. Given the growth
of Soviet military power, such a coalition approach has become a
strategic imperative if we are to preserve our collective interests at
politically acceptable cost.

But the key question is not whether a coalition strategy and posture is desirable or even affordable; it is whether the United States
and its allies can summon forth the political will to change long-
standing habits in favor of a genuine coalition approach. Can we
together achieve the wrenching change in our current nationalistic
mind- sets which is necessary to convert our alliance policies into an
alliance posture commensurate with the need? If only modestly greater
U.S. and allied defense spending is in prospect then all will depend
on whether we can collectively spend what we do get more efficiently
via more rational division of labor and vastly increased alliance cooperation — on a scale never attempted before. This in turn will
require a degree of vigorous and coherent U.S. leadership not visible
in the recent past.

Whether all this will be forthcoming, is of course, debatable. Past
omens are not auspicious. On balance, however, the advantages of
the coalition approach so far outweigh those of a primarily maritime
strategy that the most sensible course would be to try. What other
alternative do we really have? America would find it difficult to live

and prosper in a world in which we dominated the seas but our chief competitor dominated the vast resources of the Eurasian landmass.

Therefore, what should be done? First and foremost, the issues raised in this book are so central to our security that they must be made the topic of informed debate — both in the national security community and before the electorate at large. The upcoming national election campaign affords a splendid opportunity to this end. Should we continue to downplay the alliances so vital for any rational coalition approach to nonnuclear deterrence or seek to rejuvenate them? Can we afford to keep investing so much in the type of naval buildup which will tend to dictate a predominantly maritime strategy at the expense of a more balanced effort aimed at protecting our vital interests in Eurasia? Even to pose these issues clearly is to go a long way toward answering them.

Although this book is primarily an analysis of nonnuclear strategic options, the logic of its argument also has clear programmatic implications. Since nuclear stalemate dictates relying more heavily on more costly conventional capabilities, at least 5 *percent real growth* in U.S. defense spending for the next several years would be the prudent minimum. Now that their economic recovery is under way, we should also resume a consistent effort to get our allies to spend modestly more.

Given likely constraints on much higher defense spending, however, the United States must promptly assert bold leadership by advancing initiatives aimed at more rational burden-sharing and a more efficient coalition approach. This is the only way to avoid the necessity for more massive increases in defense spending than either this nation or its allies seem willing to support. We must not only persuade our allies that the gains would be worth the costs but provide tangible inducements to stimulate concrete cooperation along the lines suggested in Chapters 8 and 9.

Finally, resource constraints also dictate that the Executive Branch and Congress make wiser decisions on defense investment, particularly in costly "big ticket" programs, in the FY 1985 and subsequent budgets. For example, building two more enormously expensive carrier battle groups should not be given higher priority than spending funds on added readiness, more fast sea and airlift, and the like. If the Congress is unwilling to cancel one or both of the new carriers already funded, then older carriers should be retired. Without spelling out the details further here, suffice it to say that it is imperative to seek a better balanced strategy and posture, more balanced between readiness and modernization and with more balanced allocations to the services than the current overemphasis on a costly, carrier-heavy navy will allow.[3] We must face up to the necessity for choice.

NOTES

1. Brian Bond, *British Military Policy between the Two World Wars* (Oxford: Clarendon Press, 1980), p. 1.

2. "Maritime Strategy versus Continental Commitment?" *Orbis*, Fall 1982, p. 589.

3. For a more extended discussion of my views on future defense budget choices, see *Rethinking Defense and Conventional Forces*, pp. 46-52.

Index